Basic Microbiology for Drinking Water

Third Edition

Dennis R. Hill

American Water Works Association

Basic Microbiology for Drinking Water, Third Edition

AWWA Senior Manager of Editorial Development and Production: Gay Porter De Nileon
AWWA Senior Technical Editor/Project Manager: Martha Ripley Gray
Cover Art: Melanie Yamamoto
Production: Janice Benight Design Studio

Original line drawings, photographs, and other artwork in this volume have been composed by Dennis R. Hill, in his capacity as microbiologist for the Des Moines Water Works.

Library of Congress Cataloging-in-Publication Data

Hill, Dennis, 1951-
 Basic microbiology for drinking water / Dennis R. Hill. -- Third edition.
  pages cm
 Includes bibliographical references and index.
 ISBN 978-1-58321-981-2--ISBN 978-1-61300-258-2 (electronic) 1. Drinking water--Microbiology. 2. Drinking water--Purification. I. American Water Works Association. II. Title.
 TD384.H56 2014
 579.024'6281--dc23

 ISBN 978-1-58321-981-2; eISBN 978-1-61300-258-2

Printed in the United States of America

American Water Works Association
6666 West Quincy Avenue
Denver, CO 80235-3098
awwa.org

Contents

Figures

Tables

Preface

This book was written to help those with little or no microbiology education understand the basic concepts of the science. Customers and regulators expect drinking water plant managers and operators to understand the nature of bacteria, protists, and viruses, which were seldom discussed in the water purification field in the past. For this reason, it is important that water industry personnel become acquainted with basic microbiological concepts to better understand their jobs.

Along with its updated core information, this third edition presents the concept of building a progressive microbiology laboratory by including stream-lined tests for the assessment of aerobic endospore/*Cryptosporidium* plant performance, and for enumerations of phytoplankton (algae and cyanobacteria), paired with traditional coliform tests, chlorine contact time value considerations, and a method of effective data communication between laboratory personnel and plant operators.

Thirty additional photographs have been added to enhance an understanding of the text, and colored illustrations accompany many of the sections that discuss individual genera of microorganisms.

Also find herein a new appendix that discusses the methods to successfully decontaminate new mains.

I hope that this book successfully serves the drinking water managers, operators, laboratory personnel, and others who read it.

Acknowledgments

I thank Des Moines Water Works CEO and General Manager Bill Stowe and other administrative personnel for being supportive of my efforts in developing the microbiology laboratory and in writing this book. I believe that their support of my educational efforts will also aid other water industry personnel in their water quality endeavors.

I also thank the AWWA publishing staff, including Gay Porter De Nileon, Martha Ripley Gray, David Plank, and Melanie Yamamoto; along with proofreader Deborah J. Lynes and compositor Janice Benight, all of whom helped make this publication possible.

—DENNIS R. HILL

CHAPTER 1

Introduction

DRINKING WATER MICROBIOLOGY

One of the primary purposes of drinking water treatment is the removal of pathogenic microorganisms. A utility accomplishes this through the application of chemicals, physical filtration, and, occasionally, biological processes. A utility can most accurately determine its effectiveness of treatment by performing microbiological tests.

Chemical and physical methods used to assess basic drinking water quality, such as softening titrations, turbidity testing, particle counting, and chlorine contact time (CT) calculations are valuable, but they are not definitive substitutes for direct microbiological determinations. The correlation of the data attained from these methods with microbiological conditions is empirically derived. In other words, the chemical and physical tests are used to indirectly assess the microbiological quality of water by relating the results that they generate with data derived from previously conducted studies by research scientists.

The Total Coliform Rule developed by US Environmental Protection Agency (USEPA) has been a successful workhorse for several decades, but coliform bacteria are not always ideal surrogates for determining effective viral, protist, or phytoplankton removal. Furthermore, each group of microorganisms presents itself to treatment chemicals and physical filters in a different manner, requiring different assessment methods.

Tests such as those for direct *Cryptosporidium* oocyst detection have been too cumbersome and expensive for practical application in drinking water laboratories. Traditional phytoplankton identification and enumeration methods using Sedgwick-Rafter chambered slides or Utermöhl settling chambers require lengthy processing and microscopic analyses. However, as science has advanced, better opportunities for the determination of microbiological treatment effectiveness have been developed for drinking water utilities.

This third edition of *Basic Microbiology for Drinking Water* presents a battery of microbiological tests that could be easily adapted for daily use by larger utility laboratories, and for partial or periodical use by smaller utilities.

This battery is especially applicable for utilities that use surface water for their source water. Surface water utilities represent only 22 percent of US community water systems, but these systems serve 70 percent of the people. Application of the streamlined methods presented in this book could potentially advance treatment effectiveness assessments for utilities that serve the majority of drinking water consumers.

The microbiological test battery for a progressive laboratory (see chapter 4) would include the following:

- Traditional coliform detection methods, which primarily use coliform bacteria as surrogates for bacterial pathogens such as *Salmonella, Shigella, Campylobacter, Vibrio,* and *Yersinia,* as well as indicators of the general contamination of water systems by soil

- An aerobic endospore method modified by the author to be convenient and sensitive, which uses the endospores of aerobic soil bacteria as surrogates for *Cryptosporidium* oocysts

- Basic phytoplankton identification and enumeration using a simple, fast, and Accurate well slide with glycerol method developed by the author, which can be used to assess general source water algal and cyanobacterial conditions, as well as to assess chemical and filtration treatment effectiveness for these organisms

- Accurate chlorine contact time values calculations to ensure effective viral removal (enteroviruses, adenoviruses, reoviruses, hepatitis viruses, etc.), as well as effective bacterial, protist (excluding *Cryptosporidium*), and miscellaneous microorganism inactivation

- Effective data communication by the personnel performing the laboratory tests with the operators and managers who use the information for process control

DISEASE TRANSMISSION

Human bodies are targets for continuous attacks by numerous infectious microorganisms. Humans have survived these assaults by developing the biochemical and cellular defenses that comprise the immune system.

We can prevent challenges to the immune system and stay healthy by avoiding harmful microorganisms. This practice requires careful personal hygiene and good nutrition, as well as social structures that preserve or clean the natural environment. Systems for drinking water treatment, sewage treatment, medical services, vaccination, and epidemiological intervention must complement wise ecological practices to limit the occurrences of disease epidemics.

Advances in medicine have nearly freed some parts of the world of many diseases that have plagued humanity until recent times. Still, smallpox, a once widespread and often fatal illness, is the only disease that has been totally eliminated through technological efforts. Smallpox was caused by a virus that existed exclusively in humans. A worldwide, concerted effort was needed to isolate and cure all people of the illness. After that, no animal or other environmental reservoir existed to allow reinfection. (Some countries have kept viable strains for research and germ warfare use.)

Unfortunately, human, animal, and environmental reservoirs still harbor many other infectious organisms that cause diseases such as poliomyelitis, diphtheria, cholera, typhoid fever, and tuberculosis. These deadly threats still plague much of the world, and they could reemerge if social, economic, or educational structures deteriorate or if political indifference compromises social, medical, or environmental systems designed to protect public health.

Disease organisms (called *pathogens*) vary in their ability to cause illness. Resistance to environmental factors varies among these pathogens. They also have different abilities to induce disease in a host, which is referred to as the organism's *virulence* or *pathogenicity*. Ingestion of only a few organisms can cause disease in some cases. Other organisms must be present in greater numbers before they can overwhelm a host's immune system. Sometimes tens of thousands of viruses or bacteria are required to induce a disease.

However, a needle hole can contain 500 million individual bacteria. A speck of *Salmonella* bacteria on a piece of meat or in a raw egg could make a person very ill, possibly with fatal consequences in cases involving children or the elderly. Thousands of different waterborne pathogens may enter a host in a single gulp of river or lake water.

Virulent microorganisms are often distinguished as either true pathogens or opportunistic pathogens. Their capabilities to cause disease can sometimes vary, but this terminology helps clarify the manner in which diseases develop. True pathogens are organisms that nearly always cause disease in their hosts. Opportunistic pathogens are those that take advantage of a host weakened by other illnesses, injuries, malnutrition, or physical anomalies. In other words, these organisms are pathogenic only when they encounter an opportunity. Opportunistic pathogens are often part of the normal bacterial flora present in the body of the host.

Human skin and intestinal tracts are always colonized with bacteria. Mucous membranes (such as in the lining of the mouth) also harbor a variety of bacteria. In someone who suffers from a viral cold for a week or two, bacteria from the oral cavity may take advantage of weakened defenses to cause ear infections, bronchitis, or pneumonia. This phenomenon accounts for the apparent reemergence of a cold. In reality, the relapse is often not the virus returning, but an opportunistic pathogen starting a new invasion.

Intestinal disease organisms are often contracted when a person handles previously contaminated objects such as a doorknob, then touches his or her mouth (hand-to-mouth transmission). Sometimes infected people transmit the germs through food that they have prepared, by shaking hands, or by otherwise coming into physical contact with others. In these cases, intestinal organisms are transferred by contact with feces through a process called the fecal–oral route of transmission. Intestinal illness may also be spread by ingesting spoiled food or drinking contaminated water.

Respiratory diseases are spread by aerosols from coughs, sneezes, and speech. People also pick up pathogens when they touch contaminated objects such as doorknobs or sink faucet handles and then touch their eyes and noses.

Crowded environments with limited ventilation encourage rapid person-to-person disease transmission. Day-care centers are susceptible to disease transmission because of the close contact and lack of hygiene among the children.

Purification of drinking water is one of the most successful means of preventing the spread of disease in a society. Effective water purification eliminates waterborne pathogens, and the primary transmission mode of diseases is then shifted to the person-to-person contacts previously mentioned. However, if treatment facilities fail in their goal of purifying water, widespread epidemics may result with serious consequences. Impeccable operation of treatment plants must combine with persistent watershed management to promote public health in today's industrialized world.

Monitoring to detect all bacterial, viral, and protozoan species that cause waterborne disease would be a very time-consuming and expensive undertaking. For this reason, a certain group of bacteria has been selected by the water industry for monitoring to assess the overall fecal contamination of the water. These bacteria comprise the total coliform group, which includes *Escherichia coli* (*E. coli*) as one of its members. The other genera of the group are *Enterobacter*, *Klebsiella*, and *Citrobacter*. These organisms are found in large numbers in mammalian intestines and therefore in their feces. Coliform bacteria are primarily opportunistic pathogens that are safe to ingest in small numbers. However, when they are detected in a water sample, the water is assumed to be unsafe to drink because of the possible presence of pathogenic microorganisms.

Bacteria of the total coliform group are also found in soil and water. Of these environmental isolates, *E. coli* is the least common. The three non–*E. coli* coliform genera are still significant in that their presence indicates generally poor water quality, suggesting the presence of soil and possibly feces.

A group called *fecal coliforms* is often spoken of rather than *E. coli*. Fecal coliforms are primarily *E. coli* strains and a few strains of the other coliform bacteria that thrive at the body temperature of humans.

Bacteriophages are viruses that attack only bacteria. *Coliphages* are bacteriophages that attack coliform bacteria. Because coliphages are specifically linked to coliforms, their presence indirectly indicates potentially contaminated water. Also, because the coliphage organisms are closely related to human viruses, they are used as indicators for inactivation of viruses.

Conventional tissue cultures for human viruses are time consuming and difficult. Therefore, the US Environmental Protection Agency has chosen to suggest assaying water for coliphages to determine a utility's ability to inactivate viruses.

IMMUNOLOGY

The science of immunology studies the numerous ways that people ward off disease. The continuous struggle between host and invading microbe encourages an ever-developing immune potential. This fight encourages microorganisms to also adapt, leading to the rapid development of new strains and diseases. Pathogenic microbes may be contracted from contact with humans or animals, or from the environment. The human body offers many defenses against disease resulting from this contact. A respiratory virus first encounters viscous mucus in the nasopharynx that physically traps the organisms. Tiny hairs called *cilia* move the mucus outward, away from the lungs, to be swallowed or eliminated as nasal secretions. Coughing also helps physically clear the lungs of microbes trapped in mucus. Similarly, pathogens in the stomach are eliminated from the body by vomiting and intestinal pathogens by diarrhea. These functions are defensive responses generated by the body, not conditions caused by the disease organism.

Immunoglobulin A (IgA) is an antibody that is secreted into the mucus. This antibody may attach to and inactivate some invading organisms that are trapped there. If a virus successfully attaches to a cell in the nasal cavity, it injects its chromosomal material (deoxyribonucleic acid [DNA] or ribonucleic acid [RNA]) into the host cell. The virus chromosome then inserts itself into the human cell's chromosome and induces the manufacturing of new viruses. This process rapidly proceeds until the cell ruptures and disperses several viruses onto neighboring cells, where the infection continues and escalates.

A cellular chemical called *interferon* is secreted to prevent the infection from proceeding unchecked. Interferon helps to strengthen neighboring cells from viral attack and therefore impedes the progress of the infection.

White blood cells called *neutrophils* also contribute to the immune system's function of containing infection. They travel through the bloodstream until they encounter infected tissues, where they engulf and destroy viruses and bacteria.

Other white blood cells called *lymphocytes* detect the invading microorganisms as foreign to the body. In response, they produce antibodies, which attach to the invaders, inactivating and destroying them. Antibody production is a complex event, usually requiring a minimum of two weeks before enough are produced to effectively fight off infections. For this reason, the body must rely primarily on its early defense mechanisms previously described. Antibodies help most by warding off the recurrence of disease. In such instances, the immune system is already sensitive to the invading pathogen, allowing a more rapid antibody response.

Immune systems are poorly developed in infants, and antibody production is minimal until two years of age. Some antibodies pass from a mother's blood across the placenta and into the fetal blood system giving newborn babies a basic defense against disease. Breast milk is also rich in antibodies, so a mother who nurses her child supplies enhanced protection against disease along with nutrition.

MICROBIAL ENVIRONMENTAL DIVERSITY

Microorganisms are the quintessential diversifiers. A single cell may rapidly multiply into billions, creating a strong likelihood of mutations. The billions of gene variations involved in these mutations in turn allow for a great adaptive ability. Variations in genes increase the probability that resulting characteristics will allow organisms to survive changes in environmental conditions. Microbes, therefore, tolerate varying conditions and capitalize on new opportunities for nutrition and growth.

Recent studies of bacteria and protozoa have produced a revolutionary understanding that expands the idea of adaptive potential caused by mutations. The traditional view of microbes held that each species possessed definable characteristics and performed predictably in any one environment, except for mutant strains. Research has revealed, however, any one species may contain inactive genes, as well as active ones. When the microbe's environment changes, the inactive genes may become active, resulting in new characteristics. In other words, species need not undergo genetic change through mutation to survive. Inactive genes give organisms a stored potential to adapt when needed.

This discovery might powerfully affect the water industry, for example, when source water quality is allowed to deteriorate as a result of poor watershed management. If the change triggers inactive genetic potential, common microbes previously considered harmless may produce new pathogenic effects. A simple example would be a microbe that begins generating a toxin only when exposed to excessive nutrients (nitrates, phosphates, sewage) newly introduced to its environment. Such an event happened on the eastern US coast. A one-celled alga called *Pfiesteria* is very common in fresh and marine water. It remained harmless until communities next to marine bays allowed fertilizer runoff and human waste to pollute the water. Soon fish were infected and human swimmers developed large open lesions. Scientists showed that *Pfiesteria* had unexpectedly begun producing a potent neural toxin (in response to the pollution).

Such surprises may emerge in populations of other protozoa and bacteria, possibly precipitating outbreaks of waterborne illnesses. Water utilities must monitor this potential and carefully preserve healthy conditions in their watersheds.

The diversity and adaptability of bacterial metabolism give these microbes astounding abilities to persist in the environment. They thrive in all types of soil and widely varying conditions. Some even colonize the hot springs of Yellowstone National Park.

Their resistance to adverse conditions may be most evident in their ability to grow on and inside the bodies of human hosts. The human immune system and skin oils maintain a constant fight against bacterial invasion. Despite this actively hostile environment, some invading pathogens still succeed in infecting hosts. Water industry personnel must remember the persistence of microbes and maintain constant vigilance.

The sizes of microorganisms vary as greatly within the microscopic world as they do for large animals in the macroscopic world. If viruses were enlarged to the size of a pinhead, a protist such as a *Paramecium* would be the size of a whale.

This diversity builds an entire microcosm unseen by people but which affects our everyday lives.

Figure 1-1 shows models of microorganisms side-by-side and scaled to their relative sizes. Furthermore, if one were to just select a single group of microorganisms, such as bacteria or protists, it would be apparent that a great size variation exists within that selected group.

Figure 1-1   Microorganisms size comparison. On this scale, if the bacterial endospore model were the size of a chicken egg, a *Paramecium* would be the size of a whale.

ORGANIZATION OF THE BOOK

Chapter 1 of this book provides a contemporary philosophy for advancing microbiology laboratories and discusses disease transmission, immunological principles, and microbial diversity. Chapter 2 addresses the infectious bacteria most likely to pose challenges to water treatment processes, as well as bacteria that contribute to corrosion problems in the distribution system. Chapter 3 briefly summarizes the important characteristics of several other types of microorganisms: viruses, protozoa, algae, and multicellular organisms. Chapter 4 introduces analytical techniques for isolating and detecting the different groups of organisms. Chapter 5 presents a basic summary of principles of chemistry, lime softening chemistry, and disinfection techniques. Appendix A discusses scientific nomenclature, and Appendix B presents an overview of preparing new mains microbiologically for service.

CHAPTER 2

Bacteria

Bacteria are minuscule, omnipresent, and often dangerous. Their average size is $\frac{1}{1,000}$ millimeter in diameter. They colonize the human environment: homes and businesses, the soil, even the human body. Most bacteria play a significant environmental role. They degrade dead plant and animal material, are primary in the nitrogen and carbon cycles, and keep soil and natural water ecosystems balanced. Bacteria also colonize human skin, mouths, vaginas, and intestines, helping us resist infections and aiding us in food digestion and the assimilation of vitamins. Nevertheless, some bacteria can pose enormous threats to public health if conditions allow them to thrive and multiply.

BACTERIAL METABOLISM

The life processes of bacteria vary greatly from species to species. Some, called *aerobic* bacteria, such as *Pseudomonas* and *Bacillus*, require oxygen for their metabolism. Some, called *anaerobic* bacteria, such as *Clostridium* species, can grow without oxygen. Still others are able to flourish under both conditions because of their facultative metabolism. *Facultative* bacteria comprise numerous genera, including the coliform bacteria *Escherichia, Klebsiella, Enterobacter,* and *Citrobacter*.

Primitive bacteria developed with anaerobic metabolism that allowed them to grow without oxygen, which was lacking in the environment at the time. However, microbe and plant metabolism produced oxygen as an end product, and over many centuries the atmosphere became oxygen rich.

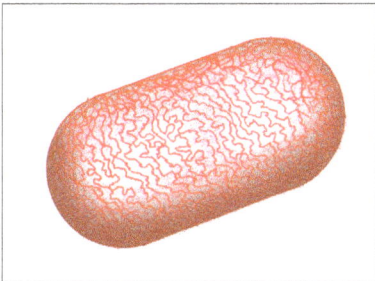

This change limited the ecological niches in which the anaerobes could grow and survive. They could not tolerate oxygen's highly oxidative effect, but they survived and remain common in low-oxygen soils, as well as in animal and human mouths and intestines.

Eventually, bacteria able to withstand contact with oxygen evolved. They developed the ability to metabolize the gas, allowing them to derive more energy from nutrients that they absorbed. Many of the bacteria found in the environment and within the bodies of animals and humans fall into this facultative group.

Strictly aerobic bacteria have also developed. Their lack of anaerobic abilities restricts their growth to oxygen-rich environments. Despite this restriction, they represent some of the most widespread water and soil organisms.

Bacteria are simple unicellular organisms. Most are free-living organisms, but a few require animal or plant hosts for survival. Bacteria absorb nutrients from their environments, excrete waste products, and secrete various toxins that help them invade tissues. Bacteria have no enclosed nucleus. Their chromosomal material is in the form of a large loop, packed into the cytoplasm of the cell.

The most common shapes of bacteria include rod, cocci (round), and spiral forms. Cellular arrangements occur singularly, in chains, and in clusters. Some species have one to numerous hairlike projections called *flagella* enabling the bacteria to swim, making them motile organisms.

Reactions to a special stain, called the *Gram stain*, separate two major groups of bacteria, gram-negative (Figure 2-1) and gram-positive (Figure 2-2) organisms. The latter have thick walls, made primarily of peptidoglycan, and stain purple.

Gram-negative bacteria have thinner walls and stain pink. A microbiologist might characterize a bacterium as a gram-negative rod, a gram-positive cocci, and so forth.

BACTERIAL TOXINS

Bacteria produce a great variety of enzymes that allow them to metabolize nutrients, and some of these chemicals also allow them to break down and invade the tissues of plants and animals. Because bacteria damage the tissues of their hosts, they are considered pathogens.

Toxic enzymes secreted by a living bacterium are called *exotoxins*. Exotoxins, most of them products of gram-positive bacteria, represent some of the most powerful toxins found in nature. Some disrupt connective tissues, while others impair neural and muscular activity or protein synthesis. Examples of these activities are seen in gas gangrene, tetanus, and botulism.

Many other less severe infective processes also result from bacterial exotoxin production by bacteria, such as inflammation of an injury.

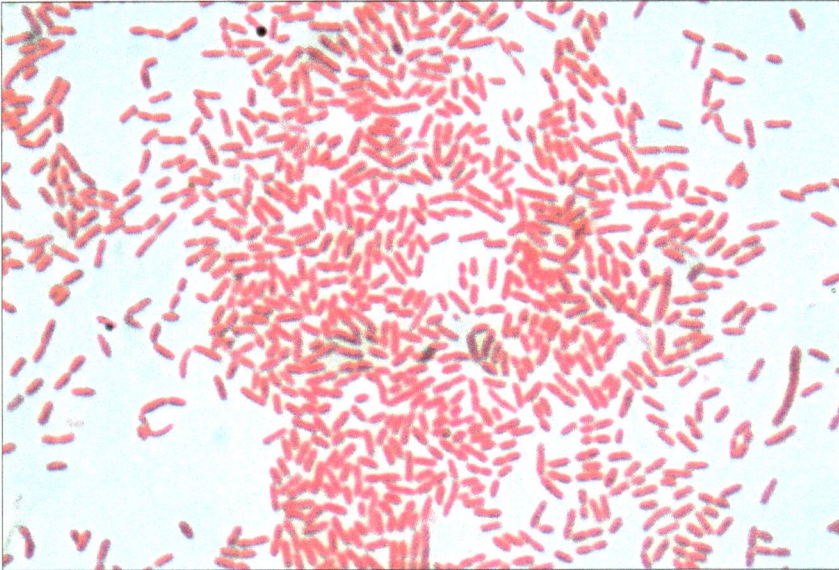

Figure 2-1   Gram-negative rods

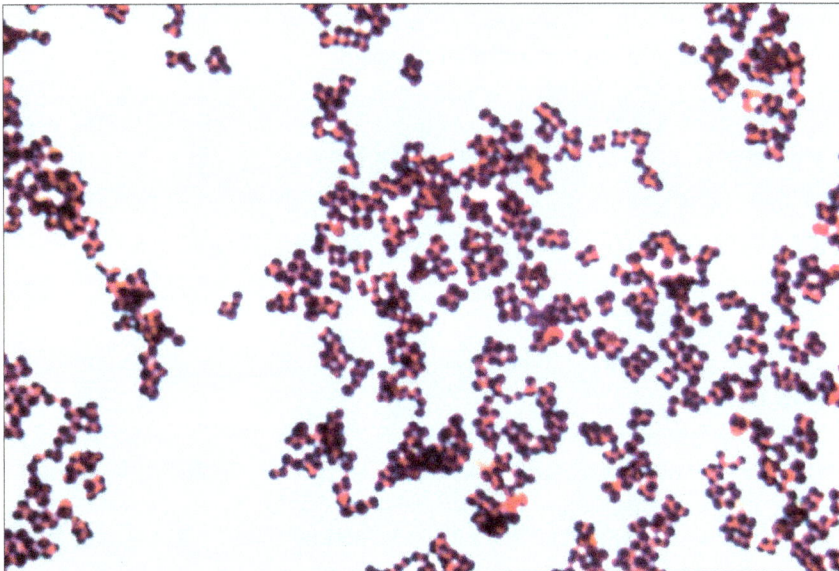

Figure 2-2   Gram-positive cocci

Another class of toxins, *endotoxins*, results only from gram-negative bacteria. These substances are components of the bacteria cell walls, so they are not released until the bacterium dies. As it disintegrates, the harmful chemicals are dispersed into the host's tissues. Endotoxin activity interferes with various metabolic processes of the host's body, possibly leading to serious consequences such as blood clotting and lowered blood pressure, leading to shock and death.

Transmission of bacterial diseases can be from person-to-person, by ingestion of contaminated food or drink, or from physical contact with animals and their waste products. Aggressively pathogenic bacteria (*true pathogens*) are transmitted in these manners.

Because bacteria can live on and in the human body as *normal flora*, the phenomenon of *opportunistic pathogenicity* is also common. This happens when harmless bacterial strains become pathogenic if the host's body is weakened by an injury, by another infectious pathogen, or by physiological disease.

People today seldom realize how they benefit from antibiotics. Without a dab of antibiotic salve applied to an infected cut, or without antibiotic pills taken for bronchitis, a urinary tract infection, an abscessed tooth, and so on, many people would suffer greatly or die. Knowledge of disease transmission, along with good hygiene, good environmental practices, and proper drinking water and wastewater treatment, are all paramount in protecting people from bacterial diseases.

Some bacterial pathogens are common in raw water sources; however, if purification systems remain functional, traditional drinking water treatment methods will effectively kill the pathogens.

BACTERIA OF INTEREST TO THE WATER INDUSTRY

Gram-Negative

Escherichia coli

E. coli is a gram-negative rod of the family Enterobacteriaceae. Most strains of *E. coli* do not produce debilitating toxins; however, clinically, they are very common opportunistic pathogens, especially when encountered in abdominal wounds and urinary tract infections. They are found in high numbers in the intestines of humans and warm-blooded animals.

A few *E. coli* strains are aggressively pathogenic or toxigenic. Enteropathogenic *E. coli* causes gastroenteritis and is usually contracted via contaminated food. *E. coli* O157:H7 is a toxigenic strain that first causes gastroenteritis and eventually may produce sufficient toxin to damage the host's kidneys, a condition known as *hemolytic uremic syndrome*. Children are especially susceptible to this strain and might die despite intensely applied medical treatment.

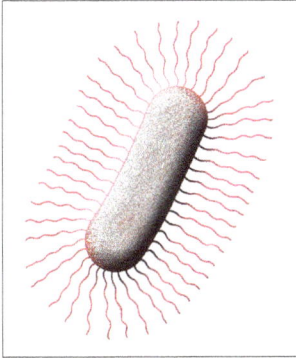

E. coli O157:H7 is found in cattle feces and consequently might contaminate processed meats. Hamburger is most notorious and should be thoroughly cooked to avoid contraction of this *E. coli* strain.

Contaminated water is also a possible source of this organism. People have contracted it from swimming pools, water parks, and fountains. Cattle yards with herds that are infected with the strain might have their wastes washed into source water streams by heavy rains or snowmelt. This would make the bacteria available for intake into a water facility.

Shigella

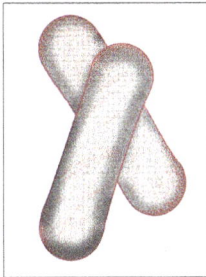

Shigella species are true pathogens. They are gram-negative rods of the Enterobacteriaceae family. All *Shigella* species cause bacillary dysentery (shigellosis), an aggressive type of bloody diarrhea.

The incidence of bacillary dysentery is low in the United States, though it flourishes in developing countries where flies, poor hygiene, and exposed food in open markets aid in its transmission. When dysentery occurs in the United States, it is likely due to the consumption of contaminated water.

Shigella dysenteriae is especially pathogenic and causes disease with a fatality rate of up to 20 percent.

Salmonella

There are numerous species and strains of *Salmonella*. They are gram-negative rods belonging to the Enterobacteriaceae family. They can be found in humans and most animal species.

Salmonella risk to humans comes from the waste of fowl, cattle, pet turtles and reptiles, cats, and unwashed vegetables. It is also transmitted from person to person, especially if those infected prepare food.

Salmonella bacteria are common inhabitants of chicken and turkey intestines. When the fowl lay eggs, the shell becomes contaminated with their feces. Though an intact egg is quite impermeable to bacterial invasion, some eggs have minute cracks, allowing *Salmonella* to enter. Once inside, the bacteria have a very nutritious environment in which to multiply. If the contaminated egg is used for consumption, it will be a potential source of food poisoning unless it is thoroughly cooked. If chicken or turkey meat is not cooked well or if the plates and utensils that touched the raw meat are used on the cooked meat, contraction of *Salmonella* becomes possible. Unpasteurized milk products may also harbor *Salmonella*.

An outbreak of *Salmonella* food poisoning occurred in peanut butter in the United States, not because peanuts pose any significant risk, but because the factories had rats that defecated into the peanut butter. The food was not heated to sufficient temperature to kill bacteria, and therefore the *Salmonella* survived to infect the product's consumers.

Rivers and lakes might contain *Salmonella* because of the many animals that carry the bacteria. This becomes especially true if manure from livestock operations enters the water.

The symptoms of *Salmonella* food poisoning—nausea, vomiting, and severe diarrhea—dramatically occur 6 to 72 hours (or longer with some strains) after eating the contaminated food. Children, the elderly, and people whose health is otherwise weakened may suffer severely and possibly die.

Salmonella, which grows in its host's intestines, can also cause septicemia. Septicemia is a dangerous invasion of the bloodstream by the organism, which can be fatal. It requires intensive treatment.

Once the illness is over, some people and animals continue to harbor this organism in their intestines without symptoms. The organism might find its way into the person's gallbladder and induce what is called a *carrier state*. *Salmonella* carriers risk contaminating food if they work in cafeterias or food-manufacturing plants.

Salmonella typhi causes typhoid fever. Despite being common worldwide, this species is rare in the United States because of careful water treatment and drinking water chlorination. Humans are its only host, which helps limit its disease. *S. typhi* has a greater tendency to cause severe septicemia than do the other *Salmonella* species. Contaminated food and water transmit the organism.

In the early 1900s, a cook, who eventually was given the name Typhoid Mary, became an identified carrier of *S. typhi*. She moved from job to job under false identities. She transmitted the disease to 51 people, causing 3 deaths over a period of 8 years. Eventually she was institutionalized until her death. Today, antibiotic treatment and the surgical removal of gallbladders cure patients of their carrier state.

Aeromonas and Plesiomonas

The genera of *Aeromonas* and *Plesiomonas* are gram-negative rods. These organisms reside primarily in aquatic environments worldwide, though *Plesiomonas* prefers warmer climates. Both genera cause human intestinal illness if consumed and must be removed from drinking water.

Vibrio cholerae

Vibrio cholerae is a curved gram-negative rod that is a true pathogen. It is not a normal part of the human intestinal flora, but rare individuals may carry it without showing symptoms of the cholera disease. This carrier state creates a potential for disease continuation in humans. The individual carrying the bacteria might at a future time contaminate a body of water from which others drink. *V. cholerae* is also able to live in brackish water and marine water.

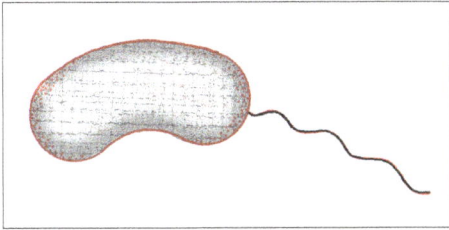

V. cholerae causes the diarrheal disease *cholera*. Cholera is a mild disease for many people that contract it; however, for some, it becomes severe and ends fatally if medical treatment is not administered. It causes epidemics and pandemics, especially in countries where sanitation is poor and sociopolitical stress occurs. This disease is frequently found in refugee camps, compounding the misfortune of the people who live there.

V. cholerae releases several toxins into the intestines. Some toxins cause extensive secretion of fluids and electrolytes (sodium and potassium) from the intestinal walls. This causes the patient to dehydrate rapidly and potentially experience fatal shock. Young children are especially susceptible.

When epidemics occur, the organism is transmitted via the fecal–oral route and through the consumption of contaminated water.

Cholera is common in Africa and Asia but epidemics of cholera have also occurred in Haiti, Pakistan, and in Central and South American countries in the last decade. In desperation, the people affected occasionally suspect political opponents of causing their disease, but undesirable social and environmental conditions are the actual precipitators of the epidemics.

Campylobacter

Campylobacter is a genus of curved gram-negative rods that is not found as normal intestinal flora in humans. It is commonly carried by various animals, including poultry, dogs, cats, sheep, and cattle. *Campylobacter* species most commonly cause gastroenteritis.

They can also cause dental disease, and systemic infections of the brain, heart, and joints accompanied by fever. *Campylobacter* gastroenteritis is the most common gastroenteritis diagnosed in the United States. It is contracted from contaminated food, milk, and water. It is seldom transmitted person to person and does not multiply in food as do many other bacterial pathogens.

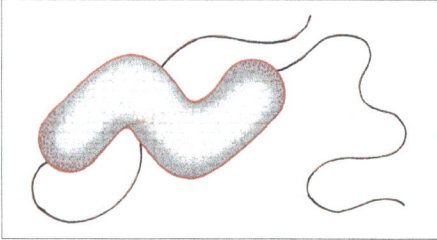

Helicobacter pylori

Helicobacter pylori is a gram-negative rod that is very similar morphologically to *Campylobacter*. It colonizes the stomachs of many people at early ages. By the age of 60, nearly 50 percent of the total population is infected.

This organism was only recently discovered, and the extent of its pathogenicity is still not known. Possible modes of transmission are oral–oral, fecal–oral, and contact with contaminated environmental sources, such as water.

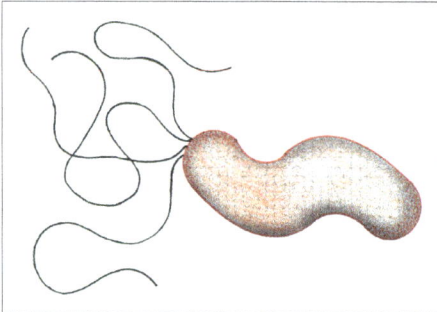

H. pylori may cause peptic ulcers and stomach cancer in people colonized by it. Discovery of its role in these illnesses has greatly changed medical attitudes, much to the benefit of patients. Stomach ulcers, once blamed on nervousness and stress, are now often cured with antibiotic treatment aimed at *H. pylori*.

In the United States, antibiotics have greatly reduced the presence of this bacterium in the stomachs of people.

This bacterium is unusual in its ability to tolerate the stomach's strongly acidic environment. It converts urea into ammonia, which helps make a less acidic niche in which it can survive.

Yersinia entercolitica

Yersinia entercolitica is a gram-negative rod. This member of the family Enterobacteriaceae is a true pathogen that causes intense gastroenteritis. Several common domestic animals host *Y. entercolitica*. Yersiniosis is contracted by eating incompletely cooked pork and consuming unpasteurized or poorly cooked dairy products. This pathogen may also be contracted by ingesting contaminated drinking water.

Legionella pneumophila

Legionella pneumophila is a thin gram-negative rod that stains very faintly with the Gram stain. It is widely distributed in the environment, especially in warm water. Rivers, lakes, and soil all harbor *L. pneumophila*, and its proclivity for warm water makes it common in air conditioners, cooling towers, and humidifiers. Whirlpools and medical equipment sometimes also contain this persistent organism. *L. pneumophila* is not transmitted by person-to-person contact. It reaches its host's lungs via aerosols, often those used in cooling equipment, humidification units, and so forth.

L. pneumophila causes the respiratory disease legionellosis, sometimes called *Legionnaires' disease*. While attending an American Legion convention at a Philadelphia hotel in 1976, several people fell victim to this severe illness, and hospital laboratories could not identify the cause. Despite the administration of antibiotics typically used to treat pneumonia, 34 of the 221 afflicted people died.

Months later, researchers at the Centers for Disease Control finally managed to isolate a bacterium later named *Legionella pneumophila*. Further study found that the organism is best treated with erythromycin, an antibiotic that is not routinely used for pneumonia treatment. This information now allows for more effective treatment of patients with legionellosis, provided they receive proper diagnosis. Laboratories can now culture and grow *L. pneumophila* and perform other tests to help identify the disease.

Pseudomonas aeruginosa

Pseudomonas aeruginosa is a small rod that stains gram-negative. It is distinctly different physiologically from the gram-negative rods of the family Enterobacteriaceae. This opportunistic pathogen is ubiquitous in nature. It can withstand and grow at a variety of temperatures, and it is resistant to several antibiotics. In particular, *P. aeruginosa* thrives in an oxygen-rich environment.

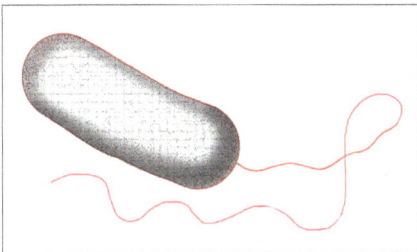

Healthy people can come into contact with *P. aeruginosa* and ingest it in small amounts without experiencing ill effects. Like other opportunistic pathogens, it requires compromised host defenses to aid it in its invasion, such as might arise from an eye injury, skin abrasion, or burn, or from prolonged contact, such as water lodged in the host's ear canal. Invasion is also aided by foreign objects present in the body, such as indwelling urinary catheters or tracheal tubes. Young patients with cystic fibrosis (a genetic lung disorder) are especially susceptible to *P. aeruginosa* and related bacteria. These children are unable to clear the bacteria from their impaired lungs, requiring frequent respiratory therapy.

Water treatment facilities normally are not concerned with *P. aeruginosa*, but the bacteria are a priority for whirlpool operators. The warm, churning, oxygenated water allows rapid chlorine dissipation and encourages *P. aeruginosa* growth. If patrons receive abrasions from a whirlpool surface, the bacteria might invade their skin, causing severe lesions. *P. aeruginosa* is resistant to many antibiotics, requiring clinical treatment with powerful and sometimes toxic antibiotics.

Mycobacterium avium-intracellulare

The species *Mycobacterium avium-intracellulare* is a member of the same genus as the bacterium that causes tuberculosis. Unlike the species *Mycobacterium tuberculosis*, *M. avium-intracellulare* is not contagious and has low pathogenicity for the average healthy person. Microscopic observation requires staining with a special acid-fast stain (Figure 2-3).

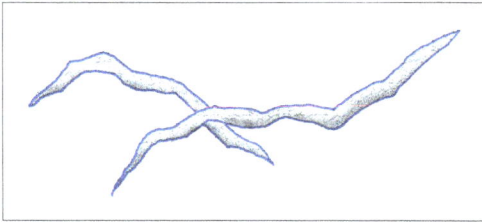

M. avium-intracellulare has been isolated from finished water in distribution systems. It is ubiquitous in the natural environment and is carried by numerous domestic animals.

Immunocompromised people suffer most from infection by this opportunistic organism. In this group, it becomes a respiratory and systemic pathogen. Disseminated *M. avium-intracellulare* disease has played a role in the deaths of 50 percent of patients with acquired immune deficiency syndrome (AIDS).

Aggressive water treatment for these bacteria is not presently encouraged. Lakes and rivers seem to be their primary reservoirs.

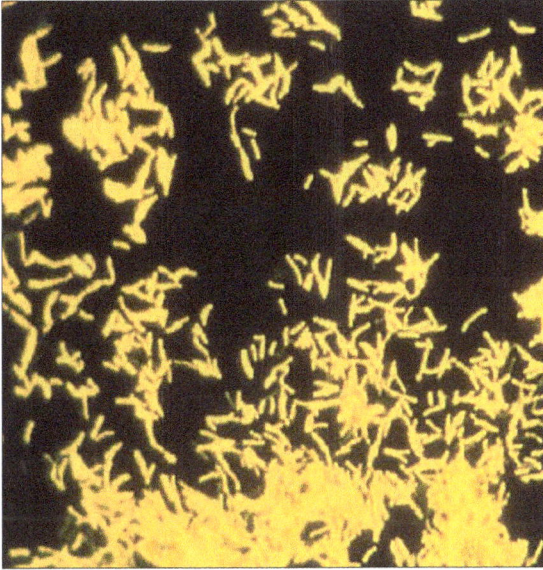

Figure 2-3   Mycobacterial stain

Gram-Positive

Staphylococcus Species

Staphylococci are gram-positive cocci. They are arranged in clusters when viewed microscopically.

S. epidermidis heavily colonizes the skin of most people. It sometimes becomes an opportunistic pathogen in conjunction with surgical stitches, skin grafts, prosthetic heart valves, and so forth.

Its only significance in the water industry comes from its common presence as a sampling faucet contaminant or as a biofilm organism. Cultures of distribution water samples collected to monitor water system quality sometimes yield large counts of S. epidermidis. This result likely comes from people touching the sampling faucet. Biofilms spoil sampling lines and deplete residual chlorine levels.

S. aureus occasionally colonizes human skin. It commonly infects wounds and sometimes becomes an invasive pathogen. Clinical treatment is performed with antibiotics, and S. aureus is becoming resistant to many of them, resulting in over 90,000 US deaths each year.

S. aureus might appear in cultures of water samples as does *S. epidermidis*. Otherwise, this species is seldom significant for the water industry. Concern is growing over *S. aureus* and other skin-infecting organisms as new attitudes toward the safety of water require more than potability.

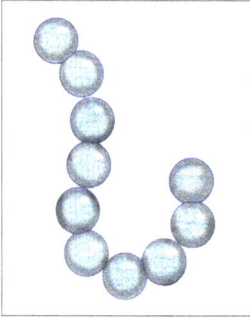

Streptococcus Species

Streptococci are gram-positive cocci that form chains. A variety of *Streptococcus* species colonize the skin, mouths, and intestines of humans, and some produce significant infections. They cause no trouble for the water industry but may be detected in sampling (Figure 2-4).

Enterococcus faecalis

Enterococcus faecalis is a gram-positive cocci that arranges in pairs and short chains. This species was once considered part of the *Streptococcus* genus. *E. faecalis* is very common in human and animal intestines. It has been used as an indicator of fecal contamination of source water, although USEPA now gives the coliform group precedence.

Clostridium perfringens

Clostridium perfringens is an anaerobic gram-positive rod and is a normal inhabitant of the human gastrointestinal tract.

C. perfringens rods are capable of encasing their chromosomes and essential organelles into tiny packages called *endospores*. These endospores

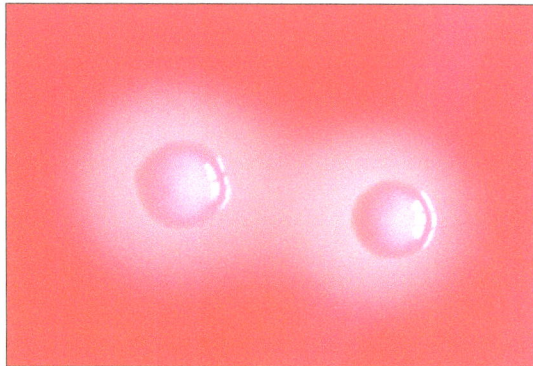

Figure 2-4   *Streptococcus* colonies on sheep blood agar plate

are very resistant to the effects of heat, desiccation, chemicals, and other environmental factors.

Multistage water treatment generally removes *C. perfringens*, and the organism poses no threat if consumed in low numbers. Nevertheless, this species can be a very aggressive opportunistic pathogen in wounds. It produces several tissue-necrotizing toxins that contribute to its infectious effects. Gas gangrene is the hallmark disease of this bacterium.

This bacterium also produces two food-poisoning toxins. With a doubling rate of 6 minutes, this species can rapidly grow to significant levels in foods held at room temperature. This ability is especially significant because its heat-resistant endospores can survive most cooking temperatures.

Because *C. perfringens* endospores have approximately the same resistance to common water treatment processes as oocysts of *Cryptosporidium parvum*, water industry personnel have used the endospores as an indicator of oocyst removal capabilities. *C. perfringens'* close relative, *Clostridium botulinum*, is the infectious agent of botulism. Similar in its temperature resistance, its toxin is extremely powerful and deadly to ingest.

Bacillus Species

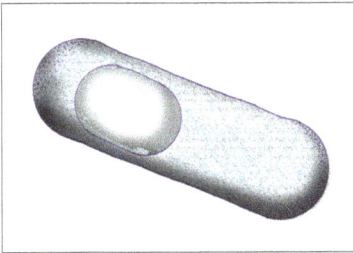

A variety of *Bacillus* species are commonly found in natural surface water and in soil. Like *Clostridium* species, they are capable of endospore production.

Most water treatment processes effectively remove or inactivate *Bacillus* species; however, some cells may survive and appear in samples. The colonies are often large and sometimes spread across the surface of culture plates, obscuring the growth of other bacterial species. *Bacillus* species are the primary soil- and water-aerobic bacteria that produce endospores. Aerobic endospores are valuable when used as assessment tools for a utility's *Cryptosporidium* oocyst removal potential. (See chapter 4.)

B. anthracis causes the disease *anthrax*, which people may contract by handling animal hides or by inhaling or ingesting the endospores. Anthrax can become a severe, even fatal infection. This species is rare in the United States, however.

Highly bred *B. anthracis* strains are sometimes propagated by military organizations as agents of biological warfare. In recent years, some individuals have also bred it to send in threatening packages to elected officials. Its potential as a terrorist weapon is worrisome, but the task of spreading it through drinking water systems would be difficult.

B. cereus and Other Bacillus Species

Food poisoning by *B. cereus* and a few other *Bacillus* species is fairly common, especially involving sandwich meats. The bacteria grow on the food and eventually produce enough toxins to cause illness. *Bacillus* species occasionally infect traumatic eye injuries. Clinical treatment includes antibiotics.

Nuisance Bacteria

Actinomycetes

The Actinomycetes group of bacteria forms thin, filamentous gram-positive rods. They are very common inhabitants of the soil, often accounting for "earthy" or musty odors through production of geosmin and other com-

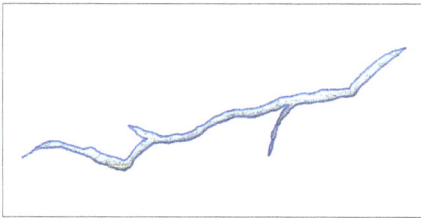

pounds. These odoriferous end products affect the aesthetic quality of drinking water, requiring treatment with powdered activated carbon (PAC) or other methods. (Geosmin is also produced by some algal species.)

Actinomycetes also colonize sand filters and sludge beds. Filter flushing and accelerated sludge bed removal helps to limit their presence and odor production. *Streptomyces* is a common genus of the Actinomycetes group.

Iron Bacteria

The iron bacteria group includes several genera: *Leptothrix, Thiobacillus, Clonothrix, Sphaerotilus, Hyphomicrobium, Caulobacter,* and *Gallionella*. These genera are widely distributed in water and soil.

They are very different from infectious bacteria such as coliforms. Iron bacteria derive energy by converting soluble (dissolved) iron into insoluble iron compounds. Some genera also precipitate manganese. When these bacteria encounter a sufficient supply of iron in well or utility water, they can produce large amounts of insoluble end products. The water becomes fouled, producing taste-and-odor problems. Human disease is not a direct consequence.

Water mains might accumulate iron bacteria and related end products. This biofilm decreases residual chlorine levels and reduces water flow. It also allows other bacteria to grow, producing pipe corrosion. Some wells can become so overwhelmed with the growth of iron bacteria that the pumps and well shafts become plugged with copious amounts of thick, slimy growth.

Water with small amounts of dissolved iron or manganese allows limited iron bacteria growth. Properly designed wells and water systems also help to discourage the growth of the bacteria. When iron bacteria are present in water rich in iron and manganese, continuous chlorination helps restrict their growth.

Sulfur Bacteria

The sulfur bacteria group includes several genera: *Desulfovibrio, Desulfotomaculum, Desulfomonas, Thiobacillus, Beggiatoa* (Figure 2-5), *Thiothrix, Chlorobium*, and *Chromatium*.

These anaerobic bacteria reduce some soluble sulfur compounds to hydrogen sulfide (H_2S) gas, giving water a rotten-egg smell.

Sulfur bacteria are often found in tubercles along with other bacteria. Tubercles are swollen, corroded spots on the interior of pipes (Figure 2-6). They can enlarge and corrode the metal until cavitation and perforation occur.

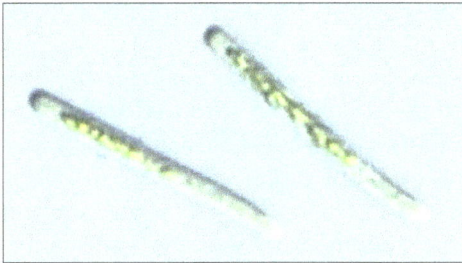

Figure 2-5   *Beggiatoa*

Figure 2-6   Pipe tubercle

Sulfur bacteria occur in combination with other genera of bacteria that produce slime, which helps protect them both from residual chlorine. Slime also helps to retain metabolic end products produced by assimilation of organic material. These end products complement the corrosion process.

Only intense chlorination can overcome the slime and tubercle barriers that sulfur bacteria form. High dissolved-oxygen levels help to prevent growth of these anaerobes.

Nitrifying Bacteria

The group of nitrifying bacteria includes *Nitrosomonas, Nitrobacter, Nitroso-vibrio, Nitrosococcus,* and *Nitrospira.* These genera are common soil bacteria that derive their energy from converting ammonia and other nitrogen compounds to other compounds, such as nitrite and nitrate, an essential step in nature's nitrogen fixation cycle. Plants then assimilate the nitrate compounds.

Ammonia applied as agricultural fertilizer is oxidized into nitrate compounds by nitrifying bacteria. If these nitrate compounds are washed or drained into waterways before being assimilated by crops, they become chemical (specifically, nutrient) contaminants in source waters.

Nitrate compounds also may be produced from otherwise organically rich soils. Erosion of these soils adds nitrates to waterways, but good land-use practices can prevent the problems.

Nitrifying bacteria can become a nuisance during water treatment. Water systems that use chloramines as disinfectants may experience breakdown of chloramines into nitrite by these organisms. Their growth might also increase disinfectant demand.

Utilities that distribute their finished water at pH 8.5 and higher will likely not have a problem with nitrate bacterial system colonization, because the bacteria do not grow in high-pH water.

Denitrifying Bacteria

Utilities in some states might encounter elevated nitrate levels in their source water. These levels sometimes exceed the USEPA drinking water allowable limit of 10 mg/L. To ensure that the finished water has a nitrate value below the limit, utilities may blend their finished water with low-nitrate water from an alternative source, such as from a deep well; they may use a charged resin nitrate removal facility; they may use membrane filtration (MF) technology; or they may pretreat the water using natural biological processes.

The latter may be accomplished if ponds or lakes are available into which the source water can be directed. If the biology of the pond or lake is well balanced, the nitrate will be denitrified or assimilated by the micro-organisms (Figure 2-7 and 2-8; see also The Nitrogen Cycle, chapter 5).

Pseudomonas species denitrify by stripping the oxygen from the nitrate molecule for respiration. The leftover gaseous nitrogen will be released to the atmosphere. Algae assimilate nitrate and use it to build amino acids. They are only about one-tenth as speedy in their nitrate usage as bacteria, but they still can be beneficial.

The nitrate level will be lowered by this biological activity, and the low-nitrate water may be blended to lower the finished water value to below 10 mg/L.

Figure 2-7   Pond used for nitrate removal

Figure 2-8   Experimental ponds showing cyanobacterial growth (left) when stagnant and minimal growth (right) when mixed

Cyanobacteria (Blue-Green Algae)

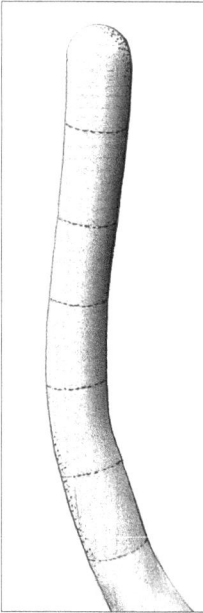

Cyanobacteria comprise a group of bacteria with many genera that have a general resemblance to algae. A few examples are *Oscillatoria, Planktothrix, Anabaena, Gleotrichia,* and *Microcystis.* Cyanobacteria perform photosynthesis as do algae; however, their cellular and physiological characteristics are that of bacteria (Figures 2-9 through 2-15).

Cyanobacteria are most notable in the water industry for the taste-and-odor problems they create. Source water may be treated with PAC to remove the odorous compounds, but cyanobacteria can colonize or proliferate in a utility's sand filters, where PAC treatment is not a feasible option. Chlorination of the filters might be required to control the problems.

Cyanobacteria propagate best in calm water and are only rarely found in some rivers with strong flow. When cyanobacteria are found in well-flowing rivers, their origin is often in the outflow of lakes. If ponds or lakes grow cyanobacteria in large quantities, enough toxins can accumulate from their metabolism to poison livestock that drink the water. Evidence suggests that humans would be sensitive to these toxins also.

Various types of cyanotoxins can be formed. Anatoxins and saxitoxins attack the nervous system; microcystins attack the liver; and cylindrospermopsins attack the liver, kidneys, and other tissues. Other cyanotoxins can also cause skin damage.

If water utilities draw their raw water from cyanobacteria-laden bodies of water, there is a chance that cyanotoxins could make their customers ill. Chlorination of water below pH 9 will destroy microcystins and cylindrospermopsins but not anatoxins. Only PAC, nanofiltration, reverse osmosis (RO) filtration, and ozone treatments effectively remove or destroy all classes of the cyanotoxins.

Cyanobacteria can sometimes evade initial treatment steps, making it to a utility's filters where they can shorten filter runs significantly. Rapid assessments of source water cyanobacteria and algae can help diagnose treatment problems. (See chapter 4.)

Furthermore, whole cells and remnants of cells that pass through treatment and into distribution systems add to the organic material that transforms into disinfection by-products (DBPs). These potentially carcinogenic compounds are regulated by the USEPA. Excessive amounts will cause a water utility to violate the minimum contaminant level set through regulation, and require that a public notice bulletin be posted.

Cyanobacterial and algal particles in distribution water can be accurately enumerated by using the well slide and glycerol method described in the chapter 4 section entitled Filter-Applied, Filter-Effluent, and Finished Water Processing.

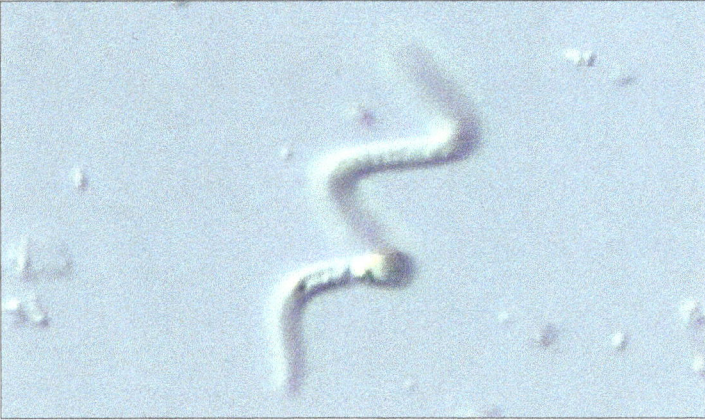

Figure 2-9   *Cylindrospermopsis*

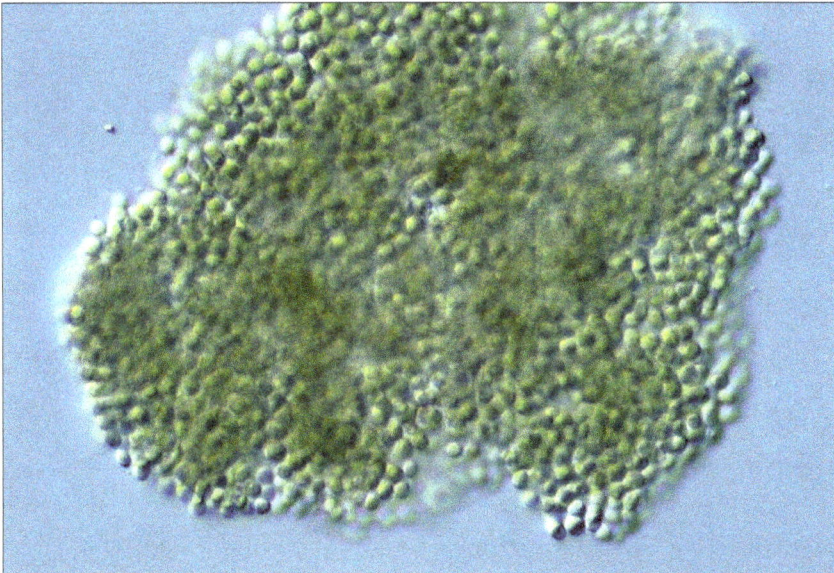

Figure 2-10   *Microcystis*

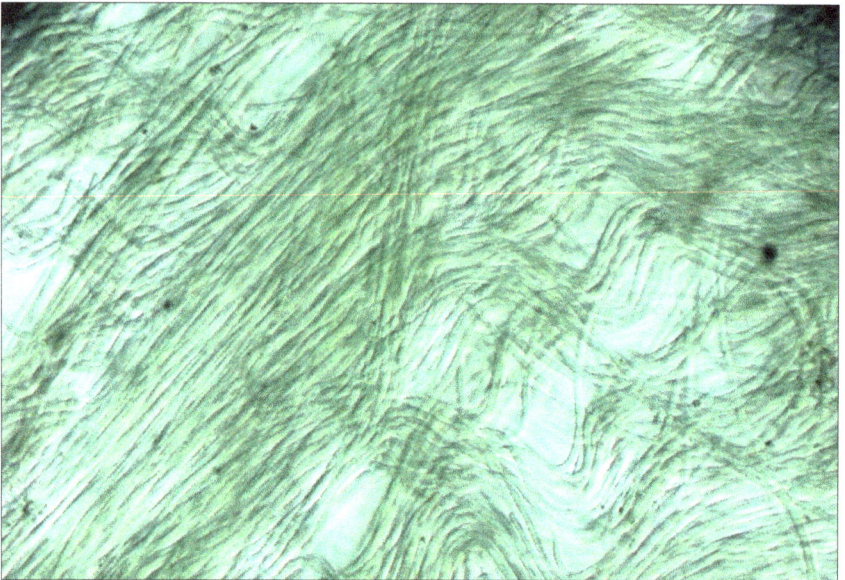

Figure 2-11   *Oscillatoria*

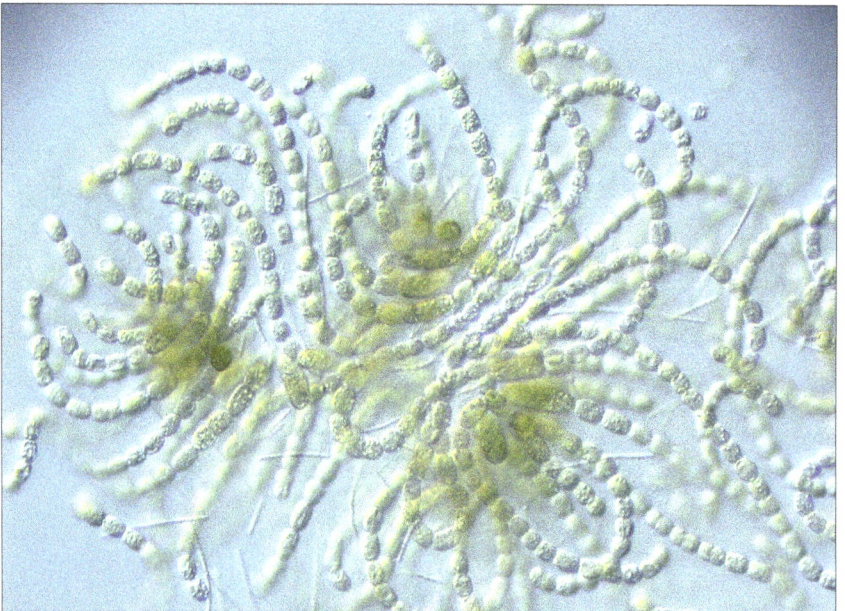

Figure 2-12   *Anabaena* with reproductive cells

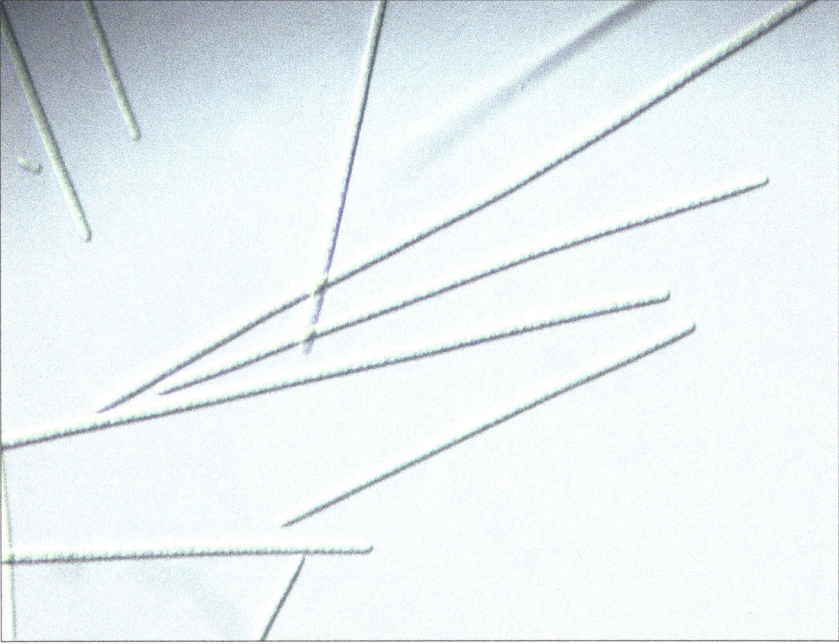

Figure 2-13   *Planktothrix*

Figure 2-14   *Aphanizomenon*

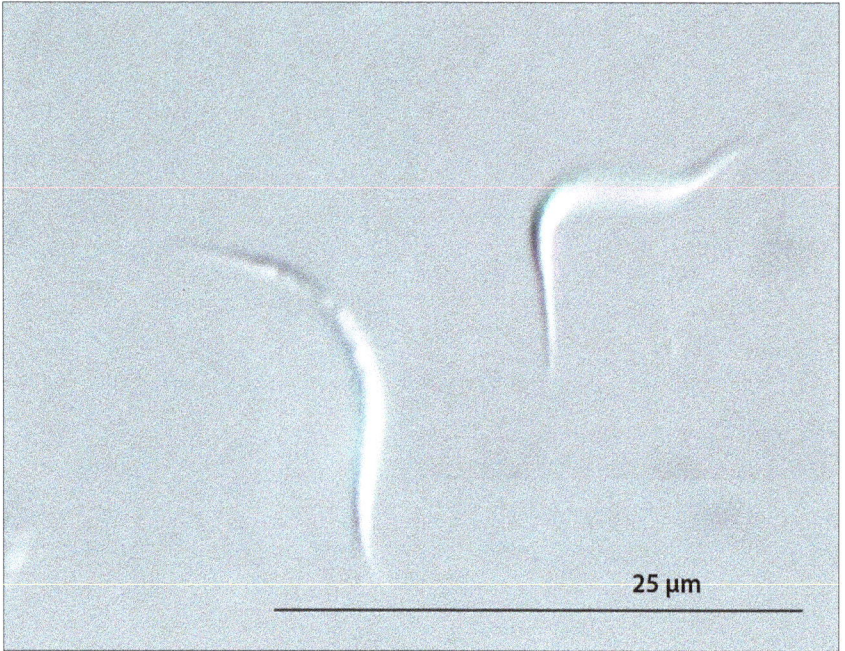

Figure 2-15   *Dactylococcopsis* cyanobacterium

CHAPTER 3

Viruses, Protists, and Other Organisms

Just as bacteria do, viruses, protists, and tiny multicellular organisms can also cause outbreaks of waterborne disease and taste-and-odor problems.

VIRUSES

Viruses are very simple, tiny life forms that do not multiply outside of living host cells. Their average size is $\frac{1}{10,000}$ millimeter in diameter. Viruses can survive in the environment for a few minutes to several hours. They are spread by hand-to-mouth transfer, by aerosol inhalation, or by ingestion.

Viruses are composed of protein packages of varying designs, all with the primary function of carrying the virus chromosome from one host cell to another. Upon attaching to the host cell, a virus injects its chromosome, which then incorporates into the host chromosome, reprogramming the cell to make viruses. Once the viruses reproduce, the cell ruptures, releasing the new viruses to scatter and infect new host cells.

These new viruses usually reinfect host cells of the same organism, but those infecting the respiratory tract may be released into the air with a sneeze or cough, or simply through speech. Hand-to-hand and fecal–oral transmissions are also possible, especially with intestinal viruses.

Viral infections are not treatable with antibiotics. They may be prevented by immunizations, careful hygiene, and effective water and sewage treatment. These methods address infection indirectly, but they have been highly successful in reducing disease, especially large-scale epidemics.

Disinfection with chlorine or ozone allows drinking water treatment facilities to easily remove most viruses from water. However, some indications suggest that viruses occasionally reactivate (become viable again) in the distribution system, despite apparently effective treatment. Utilities that practice lime softening reduce the likelihood of reactivation because the high pH enhances the killing effect. Water treatment personnel who are unsure of their utility's performance should check for viruses in water from mains distant from the point of disinfection.

Hepatitis A Virus

The hardy Hepatitis A virus causes the disease *infectious hepatitis*. It is most often transmitted by fecal–oral contact. Hepatitis A can be contracted from drinking water contaminated with sewage, or by eating food that has been handled by infected people who have not washed their hands well or who have not taken other precautions such as wearing clean food-handling gloves. Ingestion of contaminated shellfish is another common route of Hepatitis A contraction. Daycare centers occasionally are the source of outbreaks because children often exhibit poor hygiene practices.

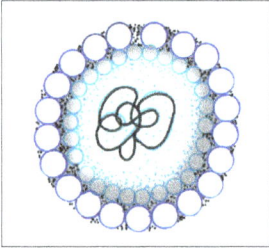

The disease ranges from mild to severe, depending on the health and age of the host. Young patients often have mild cases of the disease, while those middle-aged and elderly suffer more severe symptoms.

Hepatitis viruses attack the host's liver. Symptoms of this disease include fever, nausea, muscle aches, and vomiting; the liver may also swell and cause pain.

No effective cure for hepatitis is known, and the disease resolves itself in 4 to 8 weeks. Of the few who die from the disease, 70 percent are over 49 years old.

Enteroviruses (*Poliovirus, Coxsackievirus, Echovirus*, and Others)

The viruses of this group commonly cause intestinal disease in humans with a frequency equal to human respiratory afflictions. They may also cause generalized symptoms that affect many organs, including the brain.

A notable disease in this category is poliomyelitis, caused by the *Poliovirus*, which is now rare in the United States. An intensive nationwide immunization program has minimized the threat of transmission.

Waterborne transmission of the enteroviruses is possible, but person-to-person transmission is more common. However, as with other waterborne diseases, ineffective water treatment allows a potential for large-scale epidemics.

Gastroenteritis Viruses

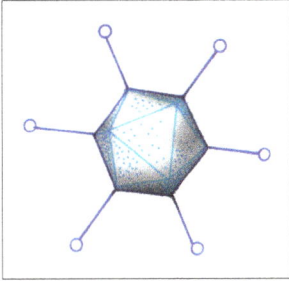

This large group of viruses is responsible for a variety of stomach and intestinal flus. It includes the *Norwalk virus, Rotavirus,* and *Adenovirus,* along with many other genera. The adenoviruses are notable because they may cause either respiratory disease or gastrointestinal disease. They are very contagious, and both waterborne and person-to-person transmissions occur.

PROTISTS

The world's organisms, both large and small, often do not conform to the nomenclature that humans wish. Scientists sometimes debate the proper classification of large animals. When classification of the myriad microorganisms is attempted, the problem to find distinct lines of division increases greatly.

For decades, it has been understood that protozoa and algae have many common traits. *Euglena* is an example of a microbe that could easily be called a protozoan, because of its similarities to a *Paramecium,* but could also be called an alga, because of its similarities to *Chlorella* and many other photosynthetic microbes.

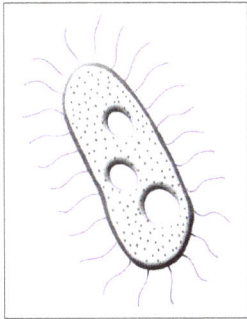

The taxonomic solution to this overlap of protozoan and algal characteristics was to place each group under one group called *Protists.* The terms *protozoa* and *algae* are still used in common communications and for general grouping of the organisms. They will also be used in this text.

Protozoan-Like Protists

The average size of protozoan-like protists is ⅟₁₀₀ millimeter in diameter.

Most protozoan-like protists are larger than bacteria. They are single-celled organisms, as are viruses and bacteria, yet they possess more complex physiologies and life cycles. In particular, a protozoan cell incorporates a nucleus that contains its chromosomal DNA.

There are a variety of forms of protozoa. In fact, they are developed to the point that generalizations about their shape and nature are difficult to make.

Paramecium

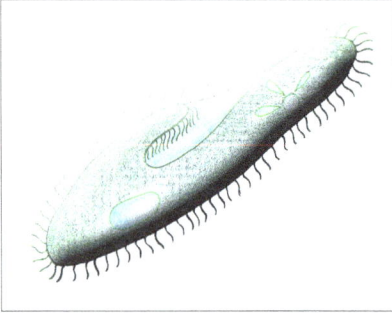

Paramecium species are present throughout nature and are one of the many protists important for ecology. They are classic examples of protists that commonly live in natural waters (Figure 3-1).

Many students encounter *Paramecia* while microscopically examining pond water. The pear-shaped cells of this genus are covered with tiny hairs called *cilia* that beat rhythmically to propel the organism through the water environment. Protozoa with cilia are called *ciliates*.

These ever-active microbes spin and cruise through the water like so many bumper cars, proliferating when nutrients and bacterial numbers are high. They and other related microbes consequently help control the balance of the ecosystem.

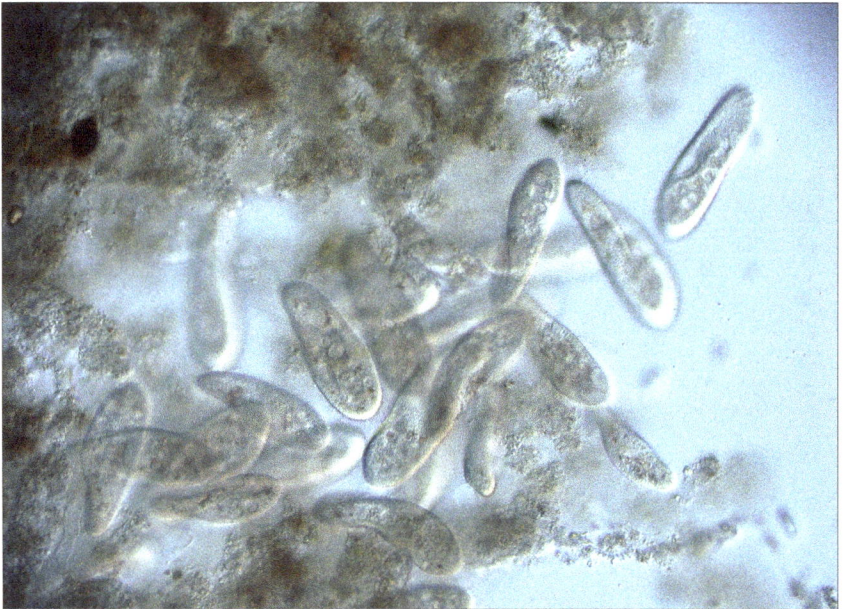

Figure 3-1   *Paramecia*

Giardia lamblia

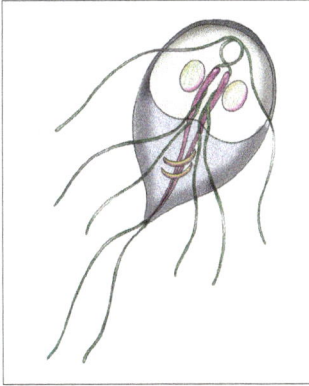

G. *lamblia* has profound importance for the water industry. It is a common intestinal parasite of humans and other mammals. It exists in two stages: the active trophozoite stage and an inactive cyst stage during which it is resistant to conditions in its environment (Figure 3-2).

Beavers are notorious for passing *Giardia*, which they harbor in their intestines, via their feces into apparently clean mountain streams. People lured by visions of nature's beauty and purity drink from the cold mountain streams and find themselves suffering from giardiasis (camper's diarrhea). This disease creates severe diarrhea, cramping, and fatigue. It is treatable with antibiotics when correctly diagnosed; however, untreated giardiasis often continues for weeks before self-resolving.

Giardia cysts are also transmitted from person to person via the fecal–oral route, an especially common problem in daycare centers where young children do not practice good hygiene.

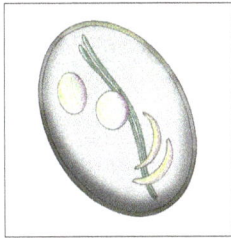

The inactive cysts are resistant to chlorine, so they might survive treatment if they pass through a facility's filtration system and if disinfectant contact time is not sufficient to overcome resistance.

Ingested *Giardia* cysts mature into trophozoites, which then multiply. Millions of organisms attach to the wall of the small intestine and cover so much of it that they interfere with nutrient absorption. Diarrhea accompanies the limited nutrient absorption, creating great fatigue and weight loss if the disease is prolonged.

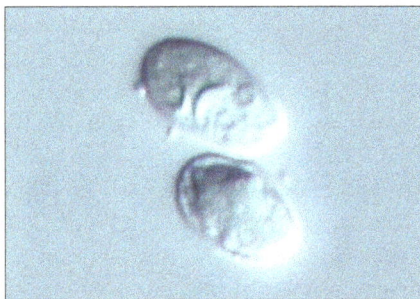

Figure 3-2 *Giardia* cysts

Cryptosporidium parvum

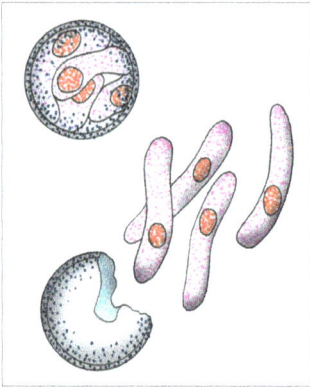

C. parvum is a species of protozoa quite different from free-living organisms, such as *Paramecia*. It is active only inside a mammalian host's small intestine where it develops into small rod-shaped sporozoites, ultimately invading the intestinal lining. The microbe develops further to the reproductive stage when it produces oocysts, each containing up to four more sporozoites. Sporozoites with thin oocyst walls are immediately released to reinfect the same host's intestinal lining. Other sporozoites with thick oocyst walls are defecated into the environment or wastewater system (Figure 3-3).

The disease of cryptosporidiosis involves severe diarrhea accompanied by intestinal cramps and fatigue. The infection lasts from a few days to a few months, depending on the strength of the host's immune system. Serological studies have found antibodies to *Cryptosporidium* in many people, indicating past infections that had not been medically attributed to cryptosporidiosis. This incidence of immunity likely accounts for the relatively mild cases that occur during epidemics.

People with compromised immune systems, such as leukemia patients, transplant recipients, infants, the elderly, and individuals with AIDS, are endangered most by the disease. This group might add up to one-fourth of the population. Antimicrobial treatment is available for treatment of cryptosporidiosis for patients with adequately strong immune systems. Person-to-person transmission is possible when proper hand washing is ignored. Also, ingestion of oocysts from lakes, streams, swimming pools, or poorly treated drinking water can cause cryptosporidiosis. Many mammals, including cattle (especially calves), hogs, and humans, are susceptible to cryptosporidiosis, allowing them to transmit the disease.

The largest waterborne epidemic of the disease in the United States occurred in Milwaukee, Wis., in 1993. Investigators believe that huge numbers of *Cryptosporidium* oocysts flowed into Lake Michigan in runoff from cattle stockyards or other flows of sanitary sewers following heavy rains. The plume of oocyst-laden water was apparently drawn into the city's drinking water facility, which uses lake water as its source.

High numbers of *Cryptosporidium* oocysts can escape coagulation with conventional treatment chemicals. A portion can then pass through

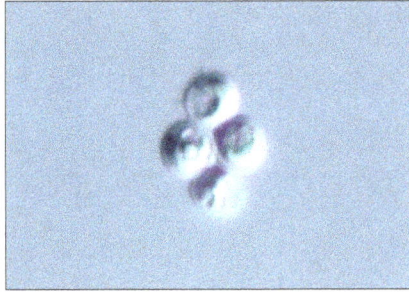

Figure 3-3   *Cryptosporidium* oocysts

granular filtration systems. The oocysts are also very resistant to the killing effects of chlorine. In Milwaukee, the infective oocysts reached the distribution system and were ingested by most of the utility's 600,000 customers. Initially, an estimated 400,000 people became ill with gastroenteritis, likely caused by *C. parvum*. This estimate has been lowered after extensive epidemiological analyses, but the significance of the incident is still profound. Up to 100 deaths occurred during the outbreak, primarily involving individuals with weakened immune systems.

Cyclospora cayetanensis

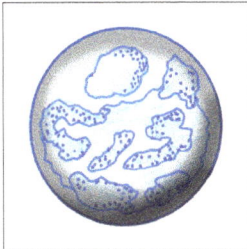

C. cayetanensis is a protist that is similar to *Cryptosporidium* in its structure and in the disease it causes. Diarrhea from the disease can last for several weeks without treatment. It has been implicated in a few waterborne disease outbreaks. It also has caused foodborne epidemics when people ingested contaminated raspberries imported to the United States from Central America. In 2013, an epidemic thought to be foodborne occurred in Iowa, where more than 70 people became ill. Ingestion of raw vegetables or fruit was the likely source. The common antibiotic Trimethoprim/Sulfamethoxazole is effective for treatment.

Microsporidia Group

This group of protozoa—primarily the genera *Encephalitozoon, Nosema, Vittaforma, Pleistophora, Enterocytozoon*, and *Microsporidium*—infects both animals and humans. These parasites invade the intestinal lining where they proliferate into high numbers, causing the host to experience intestinal illness.

The *Microsporidia* produce thick-walled spores that are resistant to environmental stresses. This resistance allows them to pass from infected individuals to new, distant hosts, via water or soil. Host-to-host transmission is also likely.

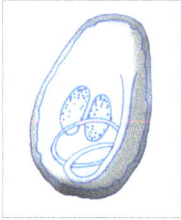

The significance of *Microsporidia* for the water industry is still not fully defined. The prevalence, treatment-resistance, and pathogenicity of these genera are presently under study.

Detection of the tiny spores (2–3 μm in diameter) is relatively difficult. Microsporidiosis is a disease with no present antibiotic treatment.

Amoebae

Effective water treatment removes many soil and water amoebae. The most notorious are *Acanthamoeba, Balamuthia, Entamoeba,* and *Naegleria.* The flexible cell walls of these organisms allow them to "ooze" in one direction or another by sending out projections called *pseudopodia.* The rest of the cell's cytoplasm follows into the pseudopodia until the entire cell has been transported. Amoebae make cysts that are resistant to environmental stresses such as drying. This helps the species survive and reach new niches (Figure 3-4).

Entamoeba histolytica is an intestinal pathogen. This invasive amoeba is contracted by person-to-person contact or by ingestion of water contaminated with feces or foods irrigated with contaminated water. When treated wastewater is used to irrigate vegetable crops, such as lettuce, cysts become trapped in the plants' crevasses. If the vegetables are not washed well, the cysts are ingested and continue their life cycle.

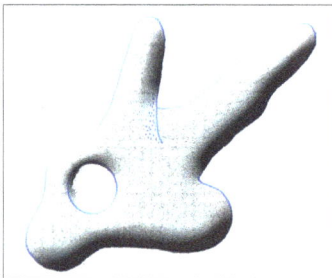

E. histolytica invades and multiplies in the host's intestinal tissue. If the disease is protracted, the amoebae travel to the liver or brain causing development of amoebic abscesses. Antimicrobial treatment becomes difficult once the infection has reached this stage.

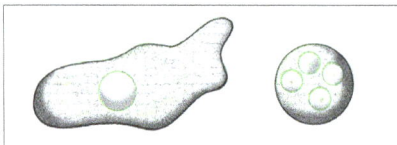

Acanthamoeba, Naegleria fowleri, and *Balamuthia* are free-living amoebae that inhabit soil, ponds, and rivers. They do not require an animal or human host to survive. *Acanthamoeba* and *Balamuthia* are rare yet particularly notorious pathogens. Each can cause severe encephalitis, a brain infection that can progress rapidly, usually with fatal results. Swimmers in ponds and lakes harboring these organisms contract them through the nasal passages. The amoebae progress to the nasal olfactory lobes and eventually to the brain.

More commonly, *Acanthamoeba* causes infections of the ears, maxillary sinuses, lungs, and especially keratitis of the eyes. Keratitis occurs when *Acanthamoeba* invades the cornea, often after contact lenses are exposed to contaminated water and then worn for an extended time. Extended exposure gives the amoeba an opportunity to create a corneal ulcer. People that prepare their own saline washes for contact lenses and those who swim while wearing their lenses have the greatest risk of infection. Corneal ulcers are painful and must be aggressively treated to prevent complete destruction of the eye.

The amoeba *N. fowleri* is a free-living soil and water organism. It can invade the human nasal chamber and eventually the brain, causing severe meningoencephalitis.

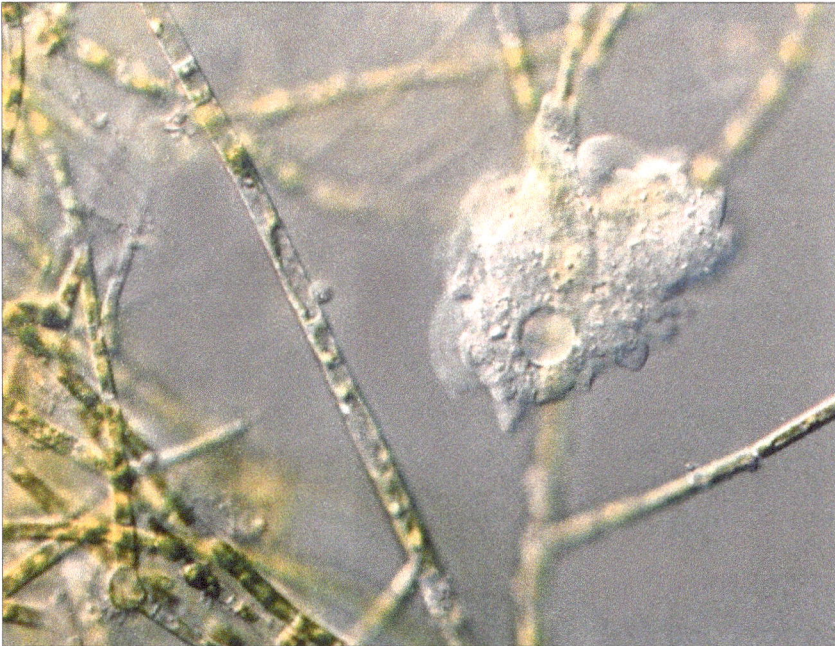

Figure 3-4   Amoeba on filamentous algae strands

Algal-Like Protists

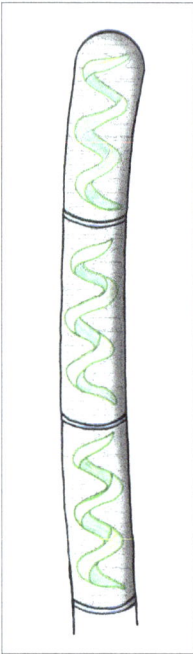

The average size of algal-like protists is 1/100 millimeter in diameter.

The surface waters of the world are filled with algal-like protists. Numerous species are included in several groups of green, yellow-green, golden-brown, Euglenoid, and Cryptomonad algae.

Numerous excellent color figures of algal groups can be found in *Standard Methods for the Examination of Water and Wastewater* (APHA, AWWA, and WEF 2012) and in AWWA Manual of Practice M57, *Algae: Source to Treatment.*

Apart from their photosynthetic process, many algae are very much like protozoa. Some exist individually as unicellular organisms (Figures 3-5 through 3-8), perhaps even with flagella for motility. Others grow in groups as colonial organisms (Figures 3-9 and 3-10), some have unique forms like the drumlike *Cyclotella* (Figure 3-11), and yet others exhibit filamentous structures (Figure 3-12 and 3-13).

Algae that remain freely suspended in water are called *planktonic algae* (Figure 3-14). Those that attach to surfaces are called *sessile algae.*

Diatoms are algae with cell walls of hard silica. They may be loosely viewed as the "glass seashells" of the microscopic world.

Dinoflagellates may threaten fish or cause "red tide" conditions that affect fishing and recreational activities. One such organism, *Pfiesteria*, produces neurotoxins that cause lesions and death in fish. Humans may also develop skin lesions, respiratory problems, and central nervous system problems if they come into significant contact with the toxins.

Dinoflagellates are common, normally harmless inhabitants of most rivers, lakes, and oceans. However, their role may change when excessive nutrients such as nitrates, phosphates, and sewage enter the water. The change in environmental conditions apparently allows dramatic proliferation of *Pfiesteria*, creating conditions for disease in fish and humans that contact the water. The Menhaden fish is the primary host of *Pfiesteria*. This fish is harvested for its oil, which is then used in cattle feed and in cosmetics.

Presently, *Pfiesteria* is no threat to drinking water facilities. Nevertheless, potential for emergence as a powerful source of disease demonstrates how environmental changes can alter an organism's role in nature in surprising and worrisome ways.

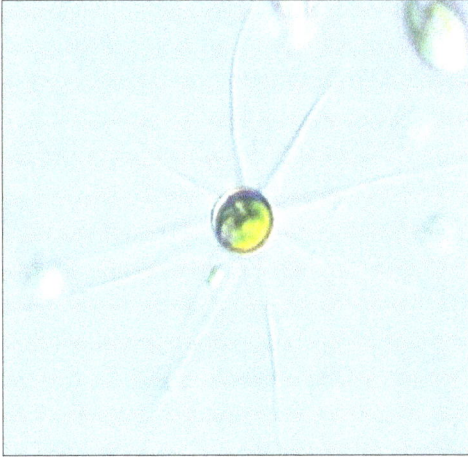

Figure 3-5   Unicellular alga

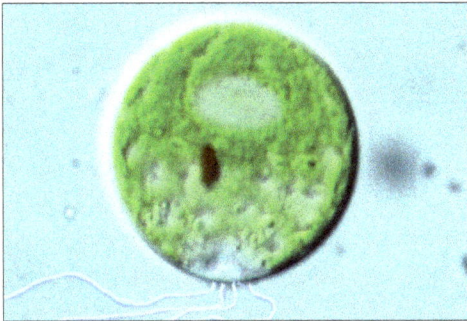

Figure 3-6   Unicellular algae

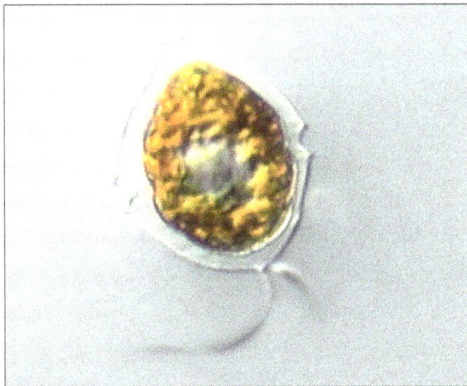

Figure 3-7   Dinoflagellate

Algae flourish during warm, sunny months when nutrients are high and ample light is available to fuel photosynthetic processes. Finished water regulations make little mention of algae, but operators at surface-water facilities usually are well aware of the potential impact of these

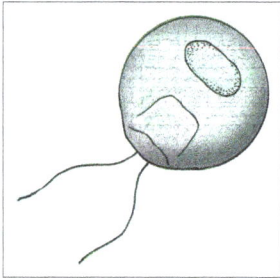

organisms on plant operations and finished water quality. Algal blooms in source waters may create taste-and-odor problems; filters may clog, requiring significantly more frequent backwashing; and finished water color problems may arise from chlorophyll that is dispersed in the water from ruptured algae cells. Whole cells and remnants of cells that pass through treatment and into distribution systems add to the organic material that transforms into disinfection by-products. These potentially carcinogenic compounds are regulated by USEPA. Excessive amounts will cause a water utility to violate the minimum contaminant level set through regulation and require that a public notice bulletin be posted.

Furthermore, strain on initial steps in multistage systems might allow increases of organic compounds derived from algae, even at the clearwell. Consequently, chlorine demand might rise.

Algae do play a beneficial ecological role. They help oxygenate water and remove numerous chemical contaminants such as phosphorus, ammonium, and nitrate compounds. They are an essential part of complex food chains, giving them a vital role in

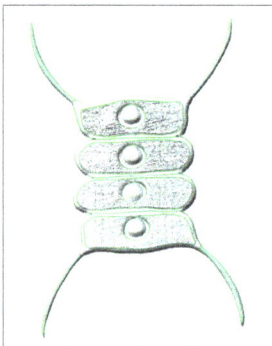

maintaining the overall health of a lake or river (Figures 3-5, 3-6, 3-14, and 3-15).

Algal blooms in a reservoir can be controlled by dispersing copper sulfate in the water. When properly administered, this practice is effective, but planning must account for the toxic effect of copper sulfate on snails and other water animals and plants that are important in maintaining a healthy reservoir. The chemical should be used at acceptable dosages.

"Blue-green" algae are actually photosynthetic bacteria called *cyanobacteria*. When people speak of "algal" toxins, they usually mean cyanotoxins. (See chapter 2, Bacteria).

Figure 3-8   Pennate diatom

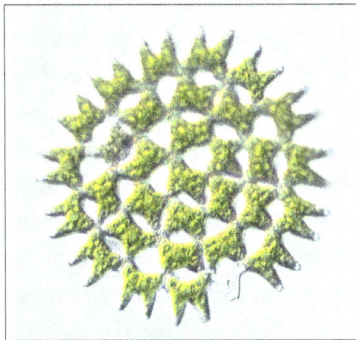

Figure 3-9   Colonial algae *Pediastrum*

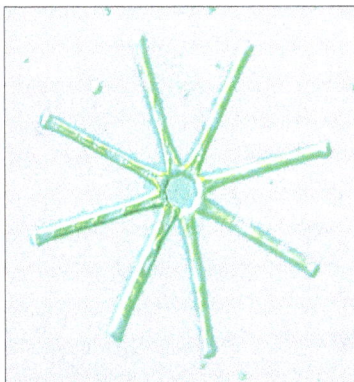

Figure 3-10   Colonial algae *Astrionella*

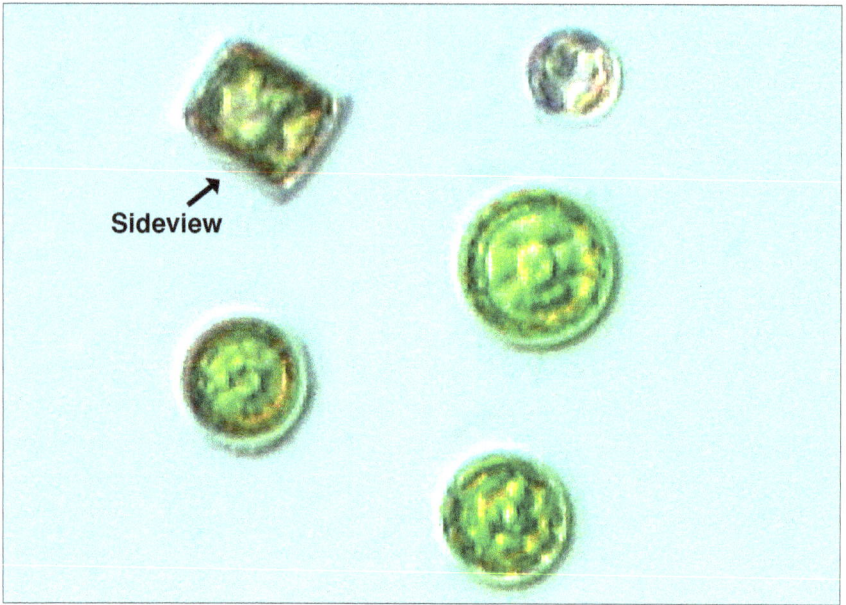

Figure 3-11   *Cyclotella* diatomaceous alga

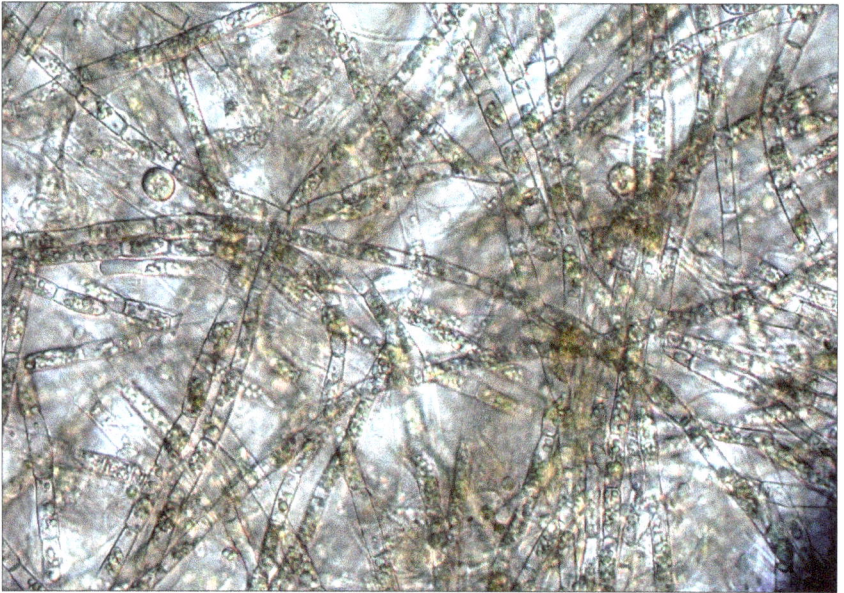

Figure 3-12   Filamentous algae

Figure 3-13   Filamentous alga *Spirogyra*

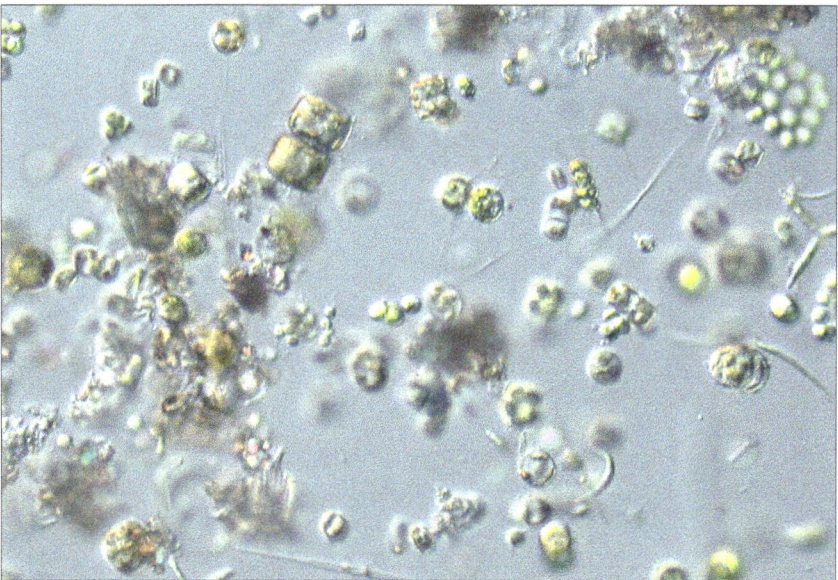

Figure 3-14   Planktonic algae

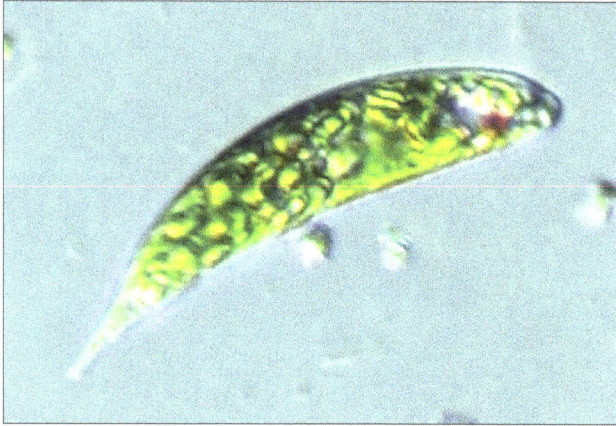

Figure 3-15   *Euglena*

MULTICELLULAR ORGANISMS

Water treatment facilities must also contend with small multicellular organisms. These organisms are not pathogenic for humans, but they might interfere with treatment processes.

Nematodes are microscopic round worms that are common in soil and water (Figure 3-16). Ideal habitats for these organisms are in the sludge blanket of a treatment basin, or in the mixed media matrix of a rapid

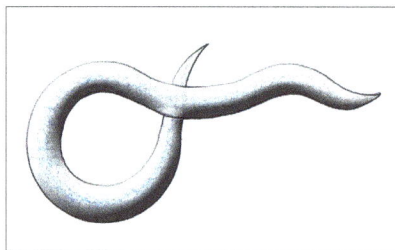

filter. Nematodes might multiply to the point that they clog the filter. Because of their impressive resistance to chlorination, they might also reach finished water. Customers with acute vision may spot the worms in their drinking glasses, compromising water aesthetics. If nematodes from source water survive treatment and pass into the finished water, they might carry along viable pathogenic bacteria that they have ingested.

Des Moines Water Works personnel have discovered that rubber lines leading to freezer ice making machines can colonize with high numbers of nematodes. This usually results in complaints about stale tasting ice.

Rotifers are microscopic creatures that flourish in water high in nutrients (Figure 3-17). They are distinguished by circles of cilia around their mouths. The movement of the cilia, resembling revolving wheels, directs

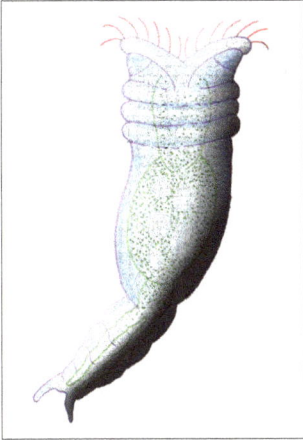

food particles into their mouths. Like tiny vacuum cleaners, rotifers can rapidly draw in thousands of bacteria for consumption, thus they play a large role along with protists in keeping the ecosystem balanced. The water industry seldom has a problem with rotifers. Small crustaceans and insect larvae might survive treatment to reach finished water at some facilities (Figure 3-18). This possibility is especially likely when water is unfiltered or when mains and filters become colonized with the organisms. These tiny organisms do not cause disease; however, their presence in drinking water is clearly undesirable. They also can disrupt filter operation, which might increase finished water turbidity.

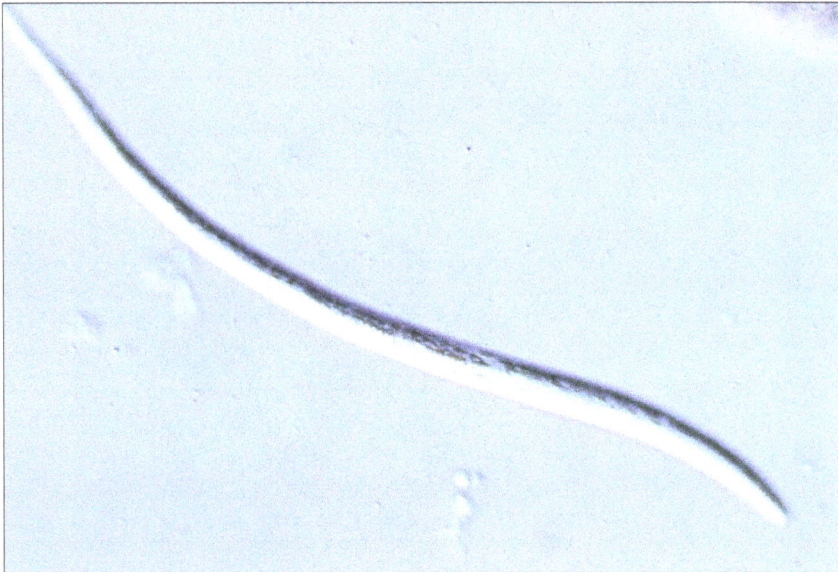

Figure 3-16   Nematode worm

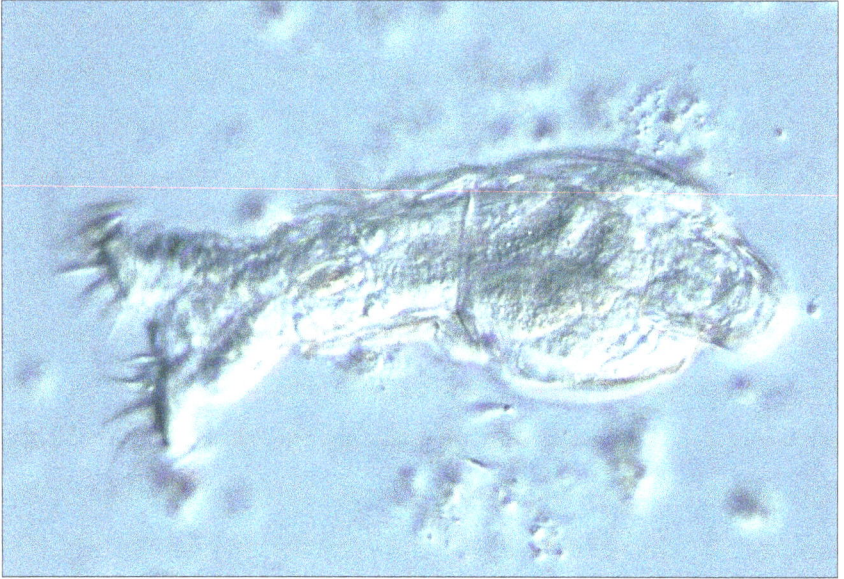

Figure 3-17   Rotifer

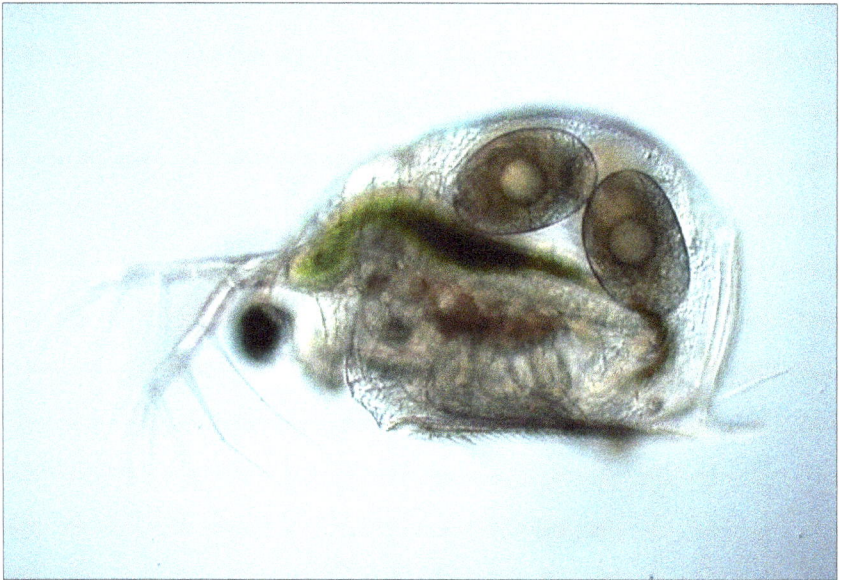

Figure 3-18   Water flea with eggs

CHAPTER 4

Laboratory Methods for Pathogen Isolation and Detection

Water treatment personnel protect the health and lives of their customers by performing a variety of laboratory tests designed to assess treatment effectiveness. This chapter introduces key microbiological methods that have been streamlined to be accurate, convenient, and inexpensive means of rapidly accomplishing this objective.

Cryptosporidium removal potentials and algal and cyanobacterial counts can now be done as a routine daily test, supplementing coliform assessments.

The federal Safe Drinking Water Act (SDWA) defines acceptable practices for sampling and analysis of drinking water. The law also grants regulatory primacy to state agencies, so their regulations take precedence, provided the states impose requirements equivalent to or more stringent than those of the SDWA. Primacy allows states to address specific needs that do not appear in other locations while ensuring effective protection for public health.

As a result, this book cannot possibly summarize all the requirements for collecting and analyzing water samples in a particular location. Instead, it gives general information that may be supplemented by specific guidance from local regulators.

Also, detailed procedures for conducting specific tests are beyond the scope of this book. More detailed information is available in *Standard Methods for the Examination of Water and Wastewater* (APHA, AWWA, and WEF 2012). Useful summaries for some procedures are available in AWWA M12, *Simplified Procedures for Water Examination*. AWWA has several other books and videos that provide additional explanation on sampling and testing (see the bibliography at the end of this book).

SAMPLING

Proper sample collection and handling are the most important consideration, and often the least discussed, in any monitoring program. The goal of any sampling procedure is to gather representative samples that accurately reflect conditions, especially the presence of pathogens, in the water being tested. Test results are only as reliable as the techniques for collecting samples.

A public water utility must gather samples according to a written plan, subject to review and revision, which specifies sampling sites, frequency, and other details. The number of samples to be taken each month is specified based on the size of the community served by the utility. If a routine sample tests positive for certain pathogens, regulations may require collection of additional samples for follow-up testing.

Samples are collected in sterile bottles and immediately dosed with sodium thiosulfate ($Na_2S_2O_3$). Sample containers must be kept closed until the moment of sample collection. Sampling should be avoided on windy and rainy days, if possible, because these conditions increase the probability of contamination. The container's lid should not be placed on any surface or held face up, which could also increase the risk of contamination. The sample should be refrigerated or iced and sent as rapidly as possible to a certified laboratory for analysis. Additional guidelines may apply, depending on conditions such as water temperature and the source of the sample.

BUILDING A PROGRESSIVE MICROBIOLOGY LABORATORY

Within this chapter you will find a battery of tests that could be employed to build a progressive microbiology laboratory. This battery is especially applicable for utilities that use surface water for their source water, but most of the methods are valuable for groundwater utilities, too. These tests could be easily adapted for daily use by larger utility laboratories, and for partial or periodical use by smaller utilities. Because microbiological purity is a primary goal of water treatment, surface water utilities serving over 50,000 customers daily are justified in having an on-site laboratory where a microbiologist performs the following procedures along with other diagnostic tests for treatment effectiveness.

A common instrument used in chemistry laboratories, such as an Ion Chromatograph, might cost $100,000 or more, and requires an $8,000-per-year service contract because of its fallibility. A microbiology laboratory equipped with an incubator, refrigerator, two water baths, centrifuge, oven, autoclave, computer, and a quality microscope with DIC (differential interference contrast) optics (very valuable), requires less than a $100,000 initial investment (Figure 4-1).

Figure 4-1   Microbiology laboratory

When staffed with a knowledgeable microbiologist who would use the progressive laboratory test battery proposed here, this laboratory would provide continuous significant information to operators and managers relating to the utility's microbiological treatment effectiveness. This would include the microbiological nature of its source water, mid-treatment water, and finished water. A utility that decides to emphasize a microbiology laboratory to the same degree or more as its chemistry laboratory, might find that far more useable information is generated relating to its treatment processes, at a lower initial cost and lower ongoing cost.

The following is an example of the value of an on-site laboratory: Des Moines Water Works uses river water as its primary raw water source. In summer and fall months when water demand is high, so is the number of phytoplankton. Certain species of algae and cyanobacteria often evade complete removal by sedimentation and flocculation. They reach the mixed media filters and shorten filter run times significantly. Rapid on-site assessments at DMWW's laboratory using fresh samples of river, treatment basin, filter effluent, and finished water greatly aid the operators in their understanding and correction of such treatment problems.

Progressive Microbiology Laboratory Battery

A progressive microbiology laboratory battery should include:

1. Traditional coliform detection methods, which primarily use coliform bacteria as surrogates for bacterial pathogens such as

Salmonella, Shigella, Campylobacter, Vibrio, Yersinia, and so on, as well as indicators of the general contamination of water systems by soil.

2. An aerobic endospore method modified by the author to be convenient and sensitive, which uses the endospores of aerobic soil bacteria as surrogates for *Cryptosporidium* oocysts.

3. Basic phytoplankton identification and enumeration using a simple, fast, and accurate well slide with glycerol method developed by the author, which can be used to assess general source water algal and cyanobacterial conditions, as well as to assess chemical and filtration treatment effectiveness for these organisms.

4. Accurate chlorine contact time values calculations to ensure effective viral removal (enteroviruses, adenoviruses, reoviruses, hepatitis viruses, etc.), as well as effective bacterial, protist (excluding *Cryptosporidium*), and miscellaneous microorganism inactivation.

5. Effective data communication by the personnel performing the laboratory tests with the operators and managers who use the information for process control.

USEPA TOTAL COLIFORM RULE

Federal and state agencies require testing of drinking water samples for total coliform bacteria. Although these organisms generally are not harmful themselves, as discussed in chapter 2, the presence of coliforms is often associated with contamination by more actively pathogenic organisms. Therefore, coliforms are accepted as an index of the microbiological safety of drinking water.

For this reason, the US Environmental Protection Agency has promulgated the Total Coliform Rule, which requires public notification and corrective action if 5 percent or more of a water utility's samples test positive for total coliforms. The Revised Total Coliform Rule was published in February 2013 with an effective date of April 2015. It has included emphasis on the detection of possible underlying problems when distribution samples grow coliform bacteria, along with the application of remedial measures, supplementing the traditional attainment of repeat samples.

Different thresholds for numbers of positive samples apply for small drinking water systems. In addition, any positive result in a total coliform test requires further sampling and retesting. If the new samples also test positive, additional testing for fecal coliforms and *Escherichia coli* is required. If these contaminants are present, they pose a serious health risk requiring a rapid response, including notification of regulators and the public.

CONVENTIONAL TESTING METHODS

Some methods for conducting bacteriological tests involve quantitative counts to determine the most probable number (MPN) of organisms in a sample. Other methods determine only the presence or absence (P/A) of organisms without attempting to quantify them. All of these methods involve combining water from the sample with liquid or solid nutrient media. The media encourage growth of coliforms while suppressing growth of other bacteria.

A solid medium called *agar* forms a gel in the testing plate. It might include a variety of ingredients, depending on the group of bacteria that the microbiologist wants to grow. Bacteria grow in colonies on the agar surface. Colonies are separate, countable groups of bacterial cells that have theoretically developed from one cell growing at an exponential pace (Figures 4-2 and 4-3).

Most media are designed to inhibit the growth of unwanted species of bacteria while nourishing the growth of the species that is being tested. The chemical compositions of some solid media cause colonies of selected species to appear a particular color. These media are called *selective media* or *differential media*. Additional biochemical tests may be performed to more accurately identify the colonies of bacteria.

Colonies may be picked off with a wire needle or loop and transferred to the special biochemical test media, where they react to chemicals during incubation. Patterns of their reaction are noted, and a database is used to identify them. One of these methods is the API 20E biochemical strip (Figure 4-4).

A liquid medium is called a *broth*. Broths of varying composition are used to grow specific types of bacteria. If organisms are present, they multiply during incubation and diffuse throughout the medium, producing a cloudy or turbid appearance.

Figure 4-5 shows three media that may be used to confirm the isolation of coliform bacteria and *E. coli*. They are lactose tryptose broth (LTB), brilliant green lactose bile broth (BGLB), and EC+MUG (4-methylumbelliferyl-β-D-glucuronide) broth. If *E. coli* is present, the latter glows blue when exposed to ultraviolet (UV) light.

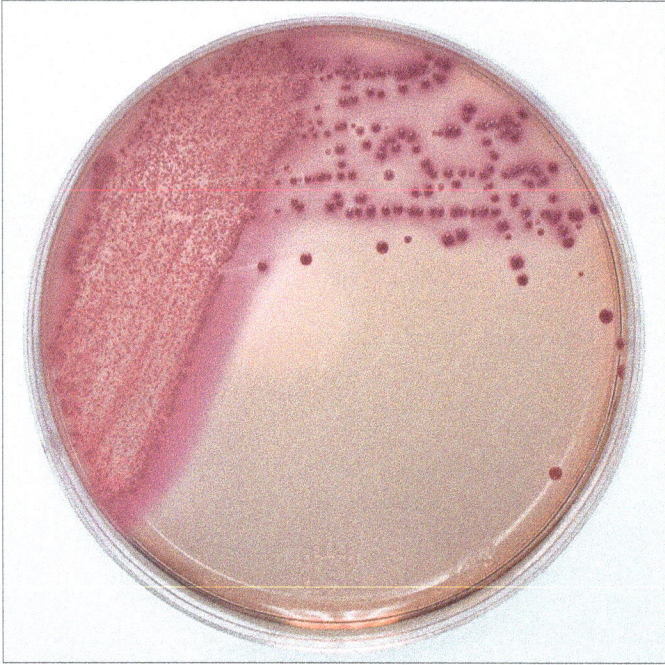

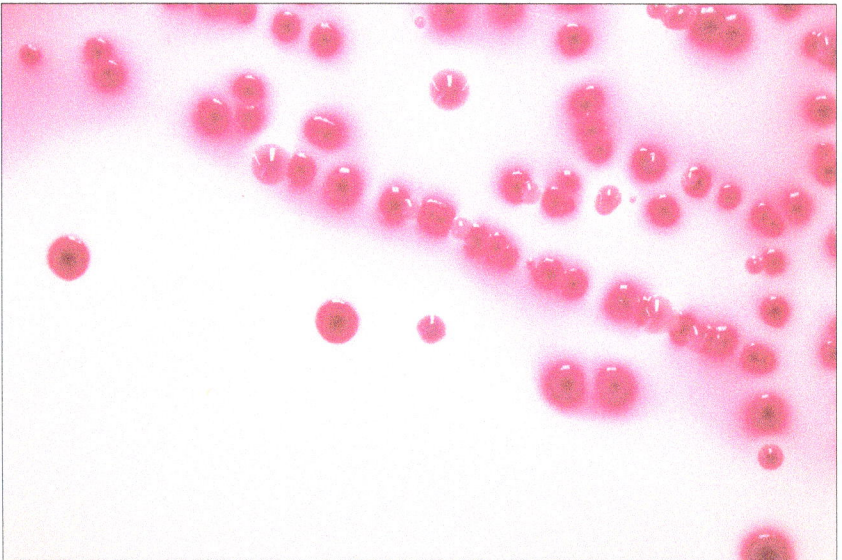

Figure 4-2   Agar plate with colonies: MacConkey agar with pink colonies
that express lactose fermentation

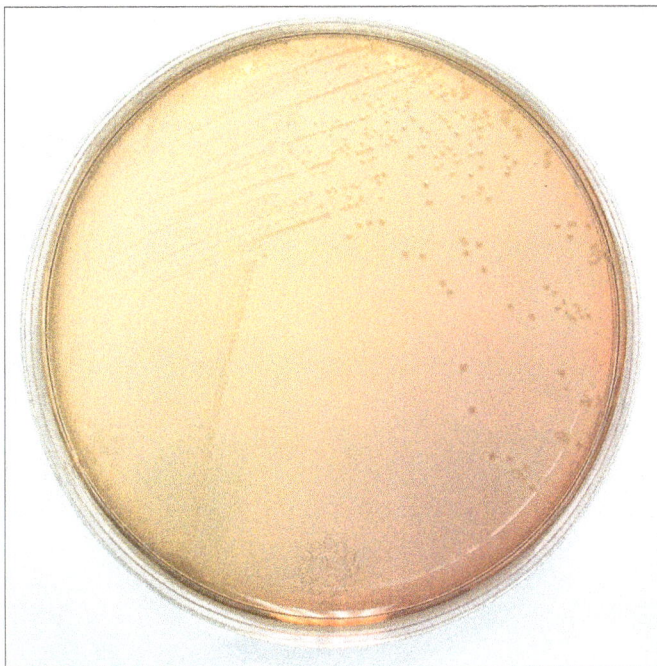

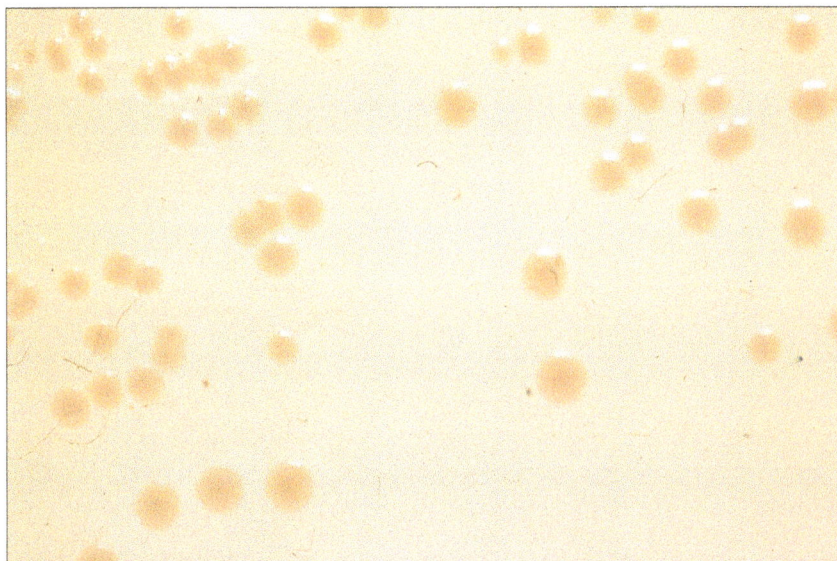

Figure 4-3   Agar plate with colonies: MacConkey agar with clear colonies that express no lactose fermentation

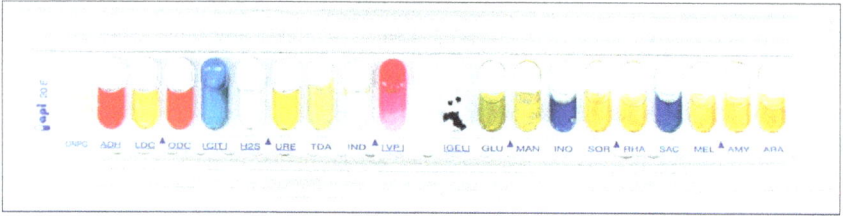

Figure 4-4   API 20E biochemical strip

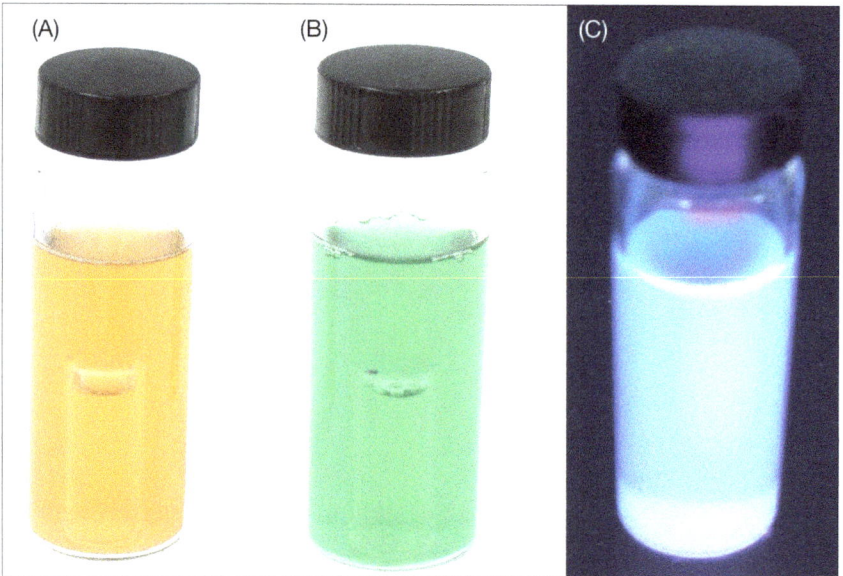

Figure 4-5   Broth tubes with growth: (A) Lactose tryptose broth; (B) Brilliant green lactose bile broth; and (C) EC+MUG broth with UV light

Heterotrophic Plate Count

Another test method called the *heterotrophic plate count* (HPC) allows for the enumeration of the bacteria present in a water sample. One of a variety of nutrient agars is prepared and kept warm enough to prevent it from solidifying. Next, 1 mL of the water sample is added to an empty, sterilized petri dish, and the warm agar is poured onto it. After a tester briefly swirls the dish to cause mixing, the agar is allowed to cool and harden. After incubation for 48 hours, the resulting colonies indicate the total count of bacteria in the sample (Figure 4-6).

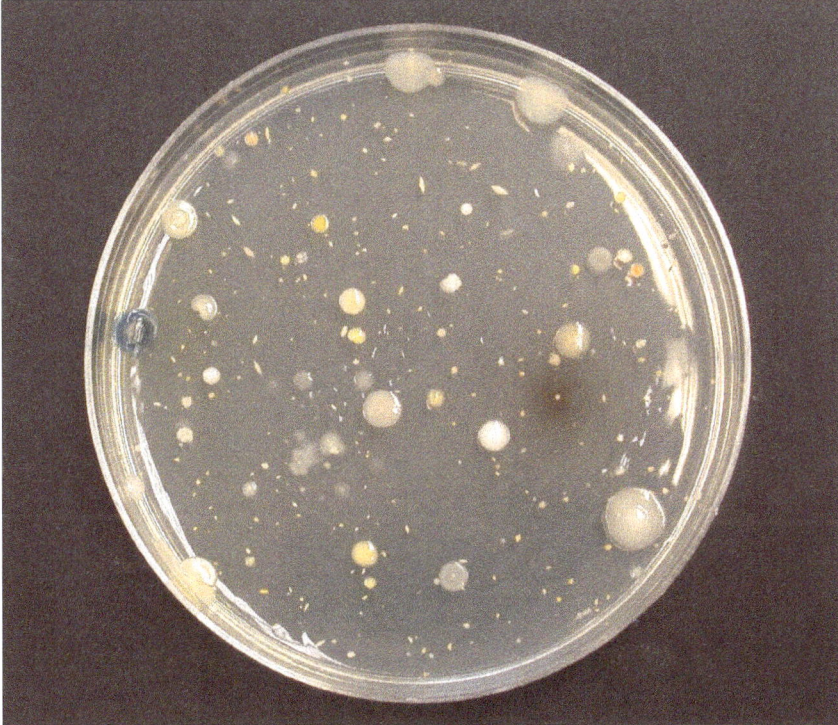

Figure 4-6   HPC plate with colonies

An alternative method exists where water may be membrane filtered, thus allowing larger volumes to be tested. In addition, an enzyme substrate method developed by IDEXX Laboratories may be used where separate wells exhibit color change.

MMO-MUG Media

MMO-MUG stands for "minimal medium ONPG with MUG." ONPG (*ortho*-nitrophenyl-β-D-galactopyranoside) is a clear compound that rapid-lactose fermenting bacteria will turn yellow. These rapid-lactose fermenters are called *coliform bacteria (Enterobacter, Citrobacter, Klebsiella,* and *E. coli).*

MUG is a compound that *E. coli* will alter so that it will fluoresce blue when exposed to UV light. The media that these compounds are added to have minimal nutrients and are balanced to grow coliform bacteria and to inhibit unimportant bacterial species.

Each of the following MMO-MUG media has been approved by USEPA for analysis of total coliform bacteria and *E. coli* in drinking water.

Colilert™ medium, which includes lactose, is added to a water sample. Coliform bacteria metabolize lactose, releasing compounds that react with a chemical in the medium to produce a yellow color after 24 hours of incubation. (There is also an 18-hour Colilert™ test medium.) Therefore, such a color change indicates the presence of any of four primary coliform genera. The test can specifically detect *E. coli* as well, because that species generates a compound that produces a bright blue fluorescence in the medium under UV light.

The Colilert™ test can also be used to generate quantitative estimates of bacteria counts. In this variation, the water–medium mixture is sealed in a cardlike tray with several wells or in multiple tubes. The number of wells or tubes that shows positive presence of coliforms is compared with an appropriate MPN index table to evaluate the extent of contamination in the sample. (See Table 9222:II in *Standard Methods for the Examination of Water and Wastewater*; also see Figures 4-7 through 4-12).

The Colisure™ media turn from a light yellow to a red or purple color when total coliform bacteria grow. The medium fluoresces blue when *E. coli* is present.

Readycult™ media turn from a light yellow to a blue-green when total coliform bacteria grow. The medium fluoresces blue when *E. coli* is present.

Colitag™ medium turns from clear to yellow if total coliform bacteria are present. The medium fluoresces blue when *E. coli* is present.

Membrane Filtration

Membrane filtration is a testing method used by drinking water and wastewater facilities to concentrate and retrieve low numbers of bacteria from relatively large amounts of water. The test can theoretically detect a single bacterium in a 100-mL sample, which amounts to the very impressive detection of 1 to 50 trillion. To concentrate the sample, the water is poured into a cylindrical funnel with a filter at its base and attached to a vacuum manifold. The vacuum draws water through the filter, capturing any bacteria present in the sample. The filter is then removed with forceps and placed into a petri dish containing a special medium, which nourishes the growth of specifically targeted organisms. The dish is incubated for 24 hours at 35°C, during which time each bacterium multiplies into a colony of billions.

Each colony theoretically represents a single organism in the original sample. However, bacteria may land on the filter in small clumps. To allow for this possibility, clumps of bacteria are counted as colony-forming units (cfu) to express the amount of bacterial recovery from a sample.

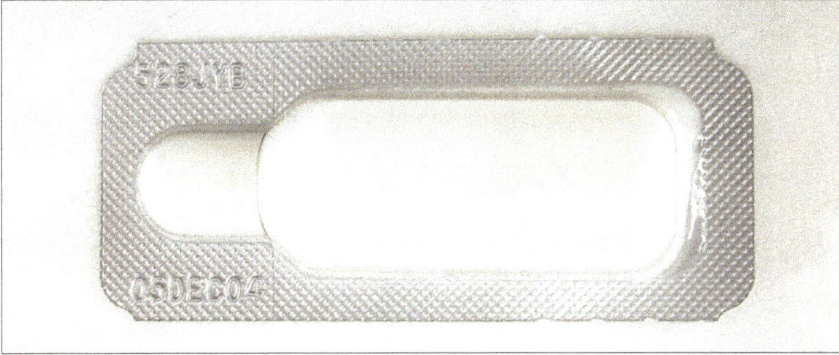

Figure 4-7   Colilert™ medium packet

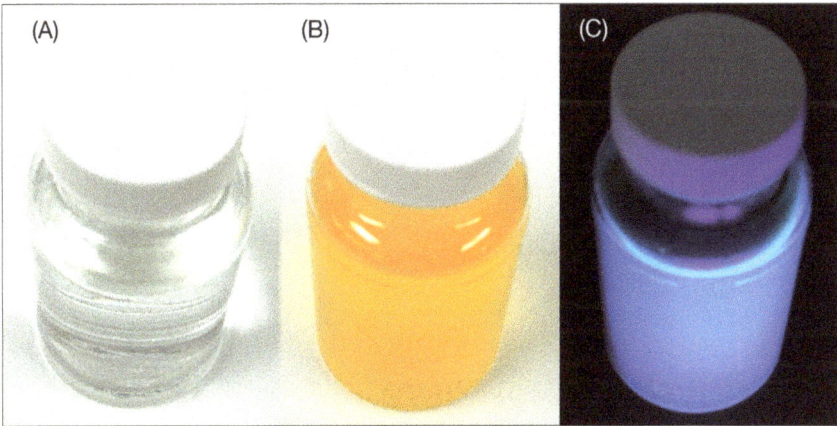

Figure 4-8   Colilert™ bottles: (A) No growth; (B) Growth of total coliforms; and (C) Growth of *E. coli* (with UV light)

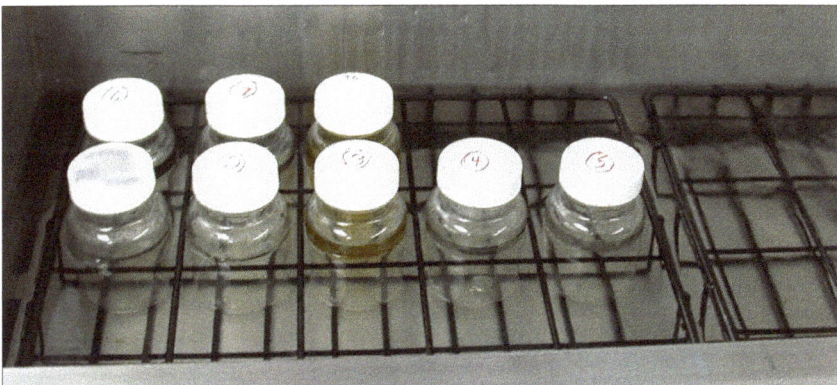

Figure 4-9   Water bath with presence/absence Colilert™ vessels

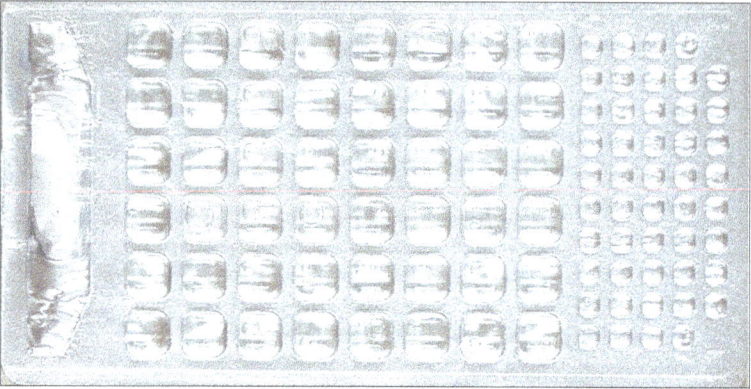

Figure 4-10   Quanti-tray 2000 with no growth

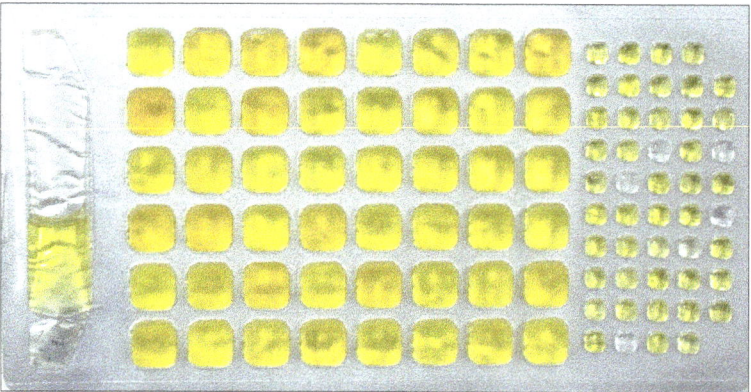

Figure 4-11   Quanti-tray 2000 with total coliform bacteria

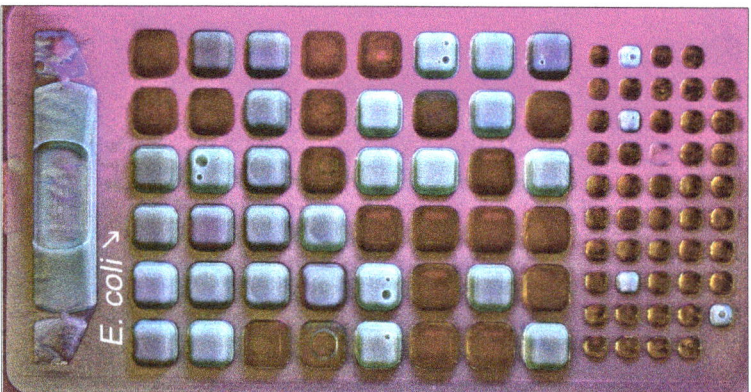

Figure 4-12   Quanti-tray 2000 with *E. coli* (fluorescence upon UV light exposure)

Media Summary

The following summarizes a variety of factors that laboratory personnel should consider when choosing media and methods for their use. These factors include the size of the laboratory, sample receipt, incubation schedule, sample type, test method strengths and weaknesses, and cost.

The two primary method formats discussed are membrane filtration and MMO-MUG media.

Membrane filtration has been the conventional method of processing water samples for coliform bacteria. It requires various laboratory apparatuses to perform the test (such as a vacuum manifold, Figure 4-15), but most of these (such as pipets and hot plates) are common to laboratories that are set up to do other water tests.

Membrane filtration is well suited for the batching of samples so that the samples are set up at a convenient time and consequently read 22 to 24 hours later. The sample is concentrated by removing essentially all of the water. The final culture plate is easily warmed, and it causes no heat–sink effect on the incubator. It also produces little biohazard waste to be disposed. The cost is also low (especially with mENDO medium) when moderate to high numbers of samples are processed. The method also allows colony counting to a circumscribed degree and provides isolated colonies for biochemical testing. Most of the samples may be declared as having no coliform growth within 22 to 24 hours, and with the newer enzyme-linked membrane filtration media, the complete result is attained once growth occurs on the plate. However, there may be intermediate colony colors on some of the newer media that laboratory personnel might find frustrating (Figures 4-13 through 4-15).

A weakness of the membrane filtration method format when using mENDO is the extended time required for additional biochemical testing of colonies when suspicious growth occurs; however, most well-kept systems only encounter occasional growth, and therefore do not have to expend more effort or time. Membrane filtration also is not ideal for small batches or for setting up samples throughout the day.

If the heterotrophic plate counts (HPCs) of a water utility's distribution samples are chronically high (>500 cfu/mL), reduced recovery of coliform bacteria might occur when using membrane filtration. This happens if the HPC organisms are noncoliform gram-negative rods that are able to grow on the otherwise highly selective membrane filtration medium. However, common distribution system heterotrophic organisms are types that normally will not grow on mENDO medium, and therefore they pose a low likelihood of interfering with coliform recovery.

Before membrane filters used for membrane microfiltration are distributed to the manufacturer's laboratory customers, they must meet the criterion of recovering greater than 90 percent of *E. coli* measured against pour plates. Accordingly, even though an individual membrane filter's bacterial recovery may exceed 90 percent, potentially 10 percent of the coliform bacteria present in a sample may remain undetected.

A filter integrity study was performed to quantify potential filter breakthrough by bacteria. In the study, no bacteria broke through the filters despite being greatly challenged, showing that the integrity of the membranes used was impressive and reliable. (See Membrane Filter Integrity Study section.)

For large laboratories that take numerous distribution samples of finished water where the cultures usually yield no coliform growth, the MF method is more convenient and the least expensive. For laboratories that only perform a few tests per week, the MF becomes costly. (See Cost Study section.)

The strengths of the MMO-MUG method center on the ease of setting up individual samples, or a few samples at a time throughout the day, and the attainment of a complete total coliform and *E. coli* result on the following day. The cost is low when only a few samples are processed, because less instrument manipulation is necessary than for membrane filtration. When several samples are set up, this cost advantage diminishes.

MMO-MUG method weaknesses center on its higher cost when processing several samples and its failure to give counts that help assess the severity of contamination incidences. Also, when acceptable plastic vessels are used for culture incubation, the samples are very slow to warm to an adequate incubation temperature. This is significant because coliform-positive samples sometimes do not turn yellow or fluoresce blue for *E. coli* until the final hour. The incubation starts once the sample has adequately warmed to temperature. This is accomplished by either prewarming the sample or by adding incubation time to the minimum required time (usually 24 hours). A sample of cold water in a plastic vessel will take approximately 30 minutes to warm to 35°C in a circulating water bath and *over 3 hours* to warm to 35°C in an air incubator.

As a compounding factor, air incubators into which several samples are placed suffer significant drops in temperature. Only large and expensive incubators are able to recover adequately following the introduction of numerous 100-mL MMO-MUG samples, especially if the samples are collected during winter months. One solution for this problem would entail the conversion from air incubators to circulating water baths for incubation. (Noncirculating or shaker/rotator baths are much less effective at sample warming than are circulating baths.)

Figure 4-13   mENDO medium for membrane filtration

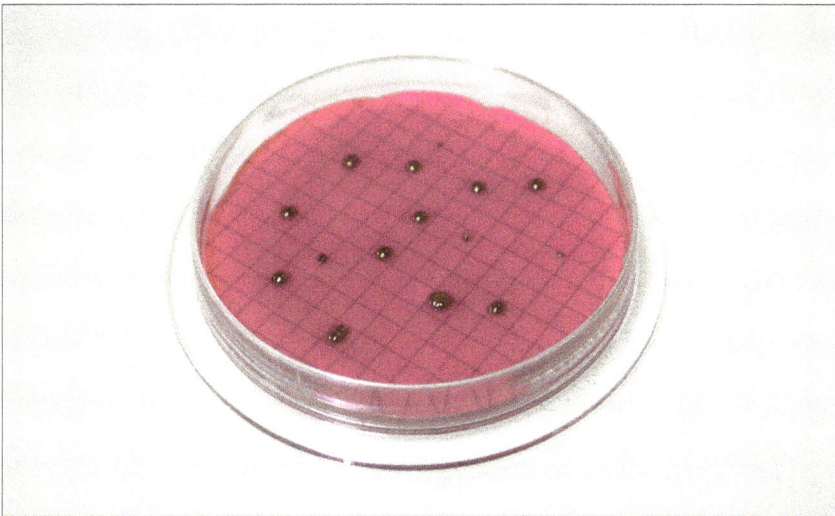

Figure 4-14   mENDO plate with sheen colonies

Figure 4-15   Vacuum manifold for membrane filtration

Colilert™ samples incubated in plastic bags can result in false-positive culture results. This outcome is due to a surprising phenomenon whereby the yellow chromogenic compound from a bag with coliform growth will transfer via air or contact to other bags in the incubator, causing them to turn yellow and therefore present falsely positive. Use of plastic bags for Colilert™ samples is no longer allowed, unless the plastic is designated for this specific use.

The strengths of the Colilert™ Quanti-tray 2000 method center on its ability to give total coliform and *E. coli* counts, in 24 hours, that fall over a wide range of recovery (0 to 2,419 MPN). It works well for raw water bacterial assessment and plant process control. In fact, it is the only method available with which to reasonably perform watershed studies, because extreme dilutions of the samples are not needed.

MMO-MUG vessels and bags, and Quanti-trays that are incubated in *large* numbers, are costly. They become cumbersome because of the incubator space that they demand, and they also cause a significant heat–sink effect on air incubators. The temperatures of the units drop and the vessels or trays are slow to warm to incubation temperature. Prewarming of the samples in water baths will eliminate this problem; however, this prewarming adds more time and manipulation to the procedure. After incubation, there remains a large amount of water-laden biohazardous waste needing proper disposal.

For small labs that perform a few samples per day/week, the MMO-MUG media methods are most convenient and least costly, because they require no media preparation or processing apparatus. They also (as with the newer MF media) do not require additional media and time when cultures grow.

The MMO-MUG media require specific and complete incubation times once the vessel's medium reaches proper temperature. This warming period adds 3 or more hours to the procedure if an air incubator is used, equaling 27 hours or more incubation time. If the vessels are incubated in a circulating water bath, only 30 minutes is added to the procedure. If the time requirements of these methods are not followed, false-positive and false-negative results are likely.

This incubation time increase constricts sample setup and interpretation windows, depending on the utility's staff number and daily work schedule. For example, if a laboratory worker's day normally ends at 4:00 p.m., a sample requiring 27 hours' incubation cannot be processed past 1:00 p.m. without creating overtime for the worker on the following day. The 18-hour test may be employed to help circumvent this problem.

Samples with a greater likelihood of growth incidence, such as new main samples, repaired main samples, and private well samples, are better undertaken using one of the MMO-MUG methods. These methods are ideal when samples are set up as they arrive throughout the day. Complete results of both total coliform and *E. coli* are often attainable in 24 hours.

Proper Incubation Time

To understand the importance of proper incubation time, especially where initial bacterial numbers are likely to be low such as with most drinking water utility samples, readers can study the growth curve in Figure 4-16. Whether considering colony formation on a petri dish medium or growth and color change within a broth medium, substantial growth does not occur until the last few hours of incubation.

The photographs of Quanti-tray wells containing 24-hour Colilert™ medium (Figure 4-17) show four wells that are insufficiently yellow to be called positive for coliform bacteria when read at 20 hours. Only one well has sufficiently turned to be called positive at 22 hours. Finally, at the full 24 hours, it is clear that three of the four wells have coliform growth.

A similar scenario would be true if vessels for presence/absence were used.

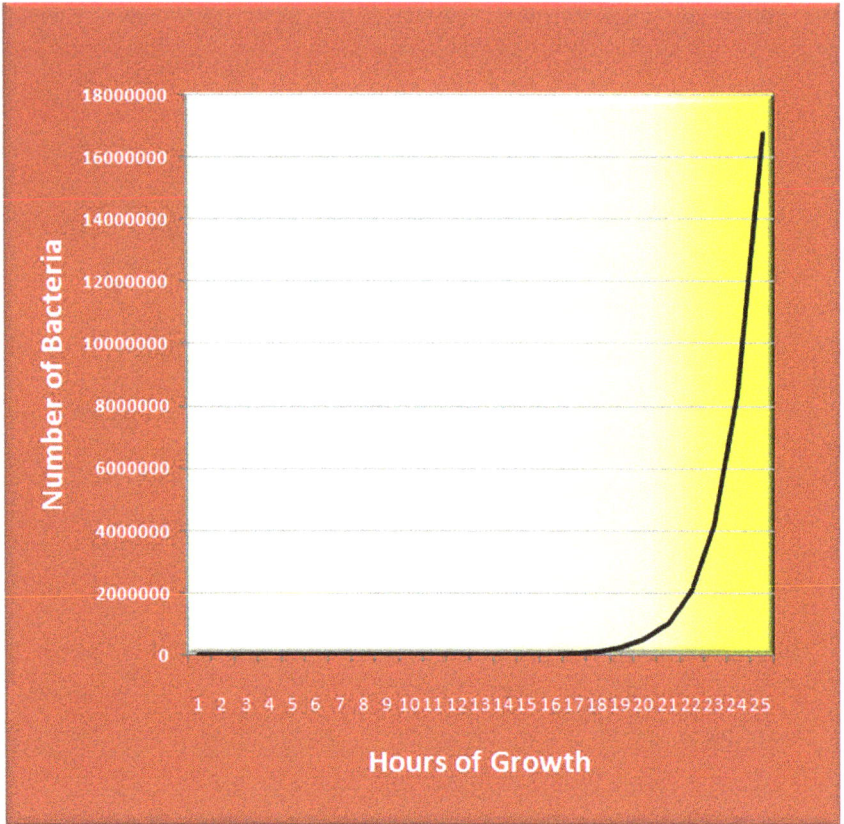

Figure 4-16   Bacterial growth curve over 24 hours and Colilert™ medium
color (yellow shading indicates color of the medium)

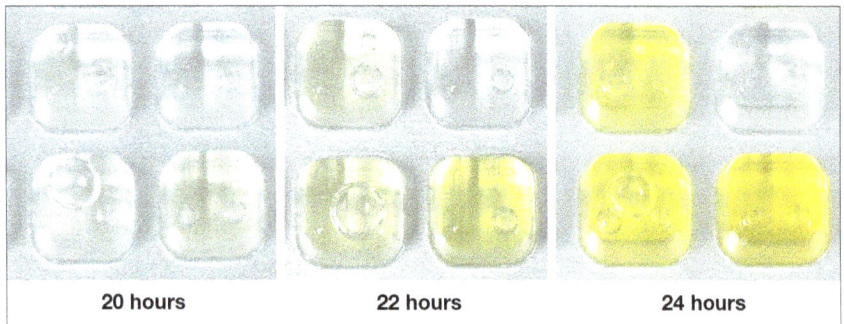

Figure 4-17   Colilert™ medium at 20, 22, and 24 hours

When Quanti-tray wells or presence/absence bottles turn yellow in less time, it is likely that the initial inoculum consisted of several bacteria, thus advancing development of the growth curve and medium color change.

Bacterial colonies growing on petri dishes exhibit a similar phenomenon, where colony appearance does not occur until the last few hours of incubation.

Some colonies might even be invisible when the plate is viewed with magnification using a dissecting microscope, if viewed a few hours before the full incubation time is met.

This circumstance once again emphasizes the importance of proper warming and incubation of samples for bacteriological study.

Specific Coliform Detection Media

Membrane filtration media are marketed in various forms. Some may be purchased as powdered media so that they may be prepared in batches. This is an inexpensive method, especially when used by larger laboratories. Others come prepared in bottles or ampules so that a few plates may be easily made by smaller laboratories.

The standard MF medium historically used by the water industry is mENDO. Colonies that grow on mENDO media and develop a metallic sheen appearance or a dark red color are potential coliform bacteria. They must be further identified using additional tests, such as the tube media LTB and BGLB. Growth with gas in these media verifies the presence of total coliform bacteria. Growth in EC+MUG broth (or a similar medium) that fluoresces blue when exposed to a UV light verifies the presence of *E. coli*.

Other MF plate media have been developed recently that allow coliform and *E. coli* verification without additional tube media. They produce different colored colonies that allow 24-hour differentiation. These media vary in the ease of interpretation, the way they are marketed, and their cost.

Each of the following membrane filtration media has been approved by USEPA for analysis of total coliform bacteria and *E. coli* in drinking water (also see Table 4-1).

- Chromocult™ medium is available in a dehydrated form. Total coliform colonies on the plate are salmon color to red, whereas *E. coli* colonies are dark blue to violet. Noncoliform colonies are light blue or turquoise. *E. coli* colonies must be confirmed by adding Kovac's reagent and looking for a pink color.

- MI™ medium is sold in one-sample ampules of prepared broth. Total coliform colonies fluoresce blue under UV light. *E. coli* colonies are blue without UV light.

- mColiBlue24™ medium is sold in one-sample ampules of pre-pared broth. Total coliform colonies are red and blue. *E. coli* colonies are blue and purple. An oxidase test may be necessary to rule out some noncoliform species.

- Coliscan MF™ medium is packaged in vials. Total coliform colonies are pink to magenta and *E. coli* colonies blue to purple. Noncoliform colonies are teal green.

- E*Colite™ medium is packaged in bags. Total coliform bacteria turn the medium blue or blue/green, and *E. coli* cause the medium to fluoresce with UV light. Samples must be held up to 48 hours to confirm *E. coli*.

Membrane Filter Integrity Study

To better quantify membrane filter integrity emphasizing the ability to capture bacteria, the author developed a study where 25 samples spiked with stock *E. coli* and 25 samples of river water were examined using membrane filtration. Pall Life Sciences filters (0.45-μm pore size, product #66191, Lot 2062006) were used along with the same company's membrane filtration funnels. Filters were randomly drawn from several boxes of 200 each.

The membranes were also studied for noticeable imperfections, such as brittleness, wrinkles, or tears, and the results were recorded.

After challenging 50 membrane filters with an average of 731 coliform bacteria, no breakthrough occurred. Furthermore, no imperfections were observed in the physical structure of the membranes. The control sample of extreme number yielded 5,170,000 *E. coli*, whereas the filtrate of this extreme control still yielded no bacterial breakthrough.

Cost Study: Membrane Filtration and MMO-MUG Media

Every lab operates differently. The choice of whether to use disposable materials versus reusable materials would be one example. Another would involve work flow, and another may pertain to available instrumentation, and so on. These differences create variables that affect the outcome of a cost study; therefore, laboratory personnel should make adjustments where appropriate.

Other factors to consider when assessing the results of a cost study are as follows:

- Costs for materials and labor that were *common* among the methods were not included in the calculations.

- Reusable bottles were included in the membrane filtration methods, where bottle clarity is not necessary. If disposable bottles are

Table 4-1   Media comparison

Medium	Membrane Filtration	MMO-MUG Liquid	Incubation Time in hours	Enzyme Substrate	Tubed Media for Confirmation	Relative Expense for <10 samples	Relative Expense for >10 samples
Bectin-Dickinson MI™	X		24	X		$$$	$$
Hach mColiBlue24™	X		24	X		$$$	$$
Micrology Labs ColiscanMF™	X		24–48	X		$$$	$
BD Difco mENDO broth MF™	X		24		X	$$$	$
Charm Sciences E*colite™		X	28			$	$$
IDEXX Colilert™		X	18 or 24	X		$	$$$
Merck Millipore Readycult™ Packets		X	24	X		$$	$$$

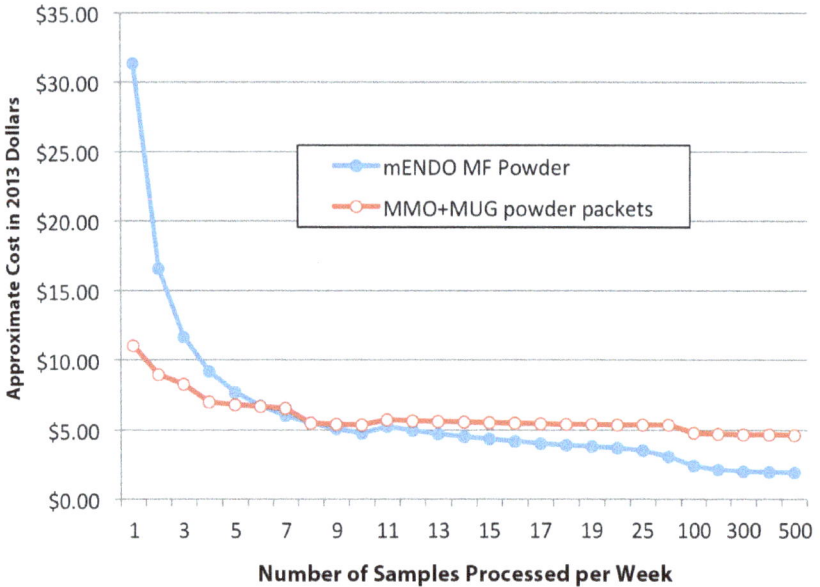

Figure 4-18   Media and method cost comparison

used for these methods, add $1.00 per test to the cost. It should be noted that use of disposable items increases ongoing purchasing expenses.

- Four hundred mL of mENDO can be made to set up 200 samples, and yet the medium cost would be $0.80 less than one packet of premade medium.

- The use of membrane filtration requires an initial investment in laboratory material that often has uses for other procedures for a moderate- to large-sized laboratory. The instruments would include a vacuum manifold and pump, hotplate, pH meter, and a balance.

- The media that use MUG require the purchase of a UV lamp.

Membrane filtration media are all very costly when a very low number of samples (fewer than 4) are processed per week. This cost diminishes as the number of samples increases. Figure 4-18 shows the cost comparison of membrane filtration and MMO-MUG methods for 1 to 500 samples per week.

MMO-MUG media (such as Colilert™ P/A and Readycult™ P/A media) initially are less expensive; however, their costs do not diminish greatly with sample number, largely because there is no time or equipment advantage to batching.

The cost of membrane filter methods drops precipitously until reaching 10 samples per week. From there, they continued to diminish in expense but less dramatically. Membrane filtration using mENDO-powdered medium soon becomes the least expensive at 9 samples per week. Once the samples reach 300 per week, the total labor and materials cost for membrane filtration with powdered mENDO medium dips below $2.00 per sample .

To address the cost of performing additional tube media tests on coliform-suspicious colonies isolated on mENDO medium made from powder, percentage values of suspicious colonies were plotted against the costs of each medium. Snapshots at 10 samples per week and at 100 samples per week were graphed.

At 10 samples per week, the cost of using mENDO powder surpasses all of the other mediums' costs when 15 percent of the cultures yield coliform-suspicious colonies. This makes the mENDO powder less cost-effective for smaller laboratories that set up relatively few samples per week and have a moderate to high incidence of suspicious colonies with which to contend.

At 100 samples per week, the cost of using mENDO powder becomes significant only when the culture yield of coliform-suspicious colonies becomes great (10 to 30 percent). This finding maintains the conclusion that mENDO powdered medium is cost-effective for use by laboratories that set up several cultures per week.

Testing for Protozoa

Water utility staff must conduct more complex analysis of water samples to determine the presence of protozoa such as *Cryptosporidium parvum* and *Giardia lamblia*. Most testing of this type involves microscopic analysis to identify specific microorganisms based on their appearance and cellular structures. Cartridge filters are available that trap protozoa in samples of water passed through them. Properly used, these devices can speed up the process of analysis and detection, but a certified lab must complete the testing.

All methods of directly testing for protozoa require concentrations of large amounts of water into manageable volumes to confine any organisms present. Any solids in the remaining sample are further concentrated using a centrifuge. This step is the most labor-intensive aspect of the testing process.

One creative method approved by USEPA is immunomagnetic separation. This procedure tags organisms with magnetite and extracts them from concentrated debris using a strong magnet. The increase in efficiency reduces the number of purification steps and thus saves time and greatly improves the recovery of organisms. (See also section on Aerobic Endospore Method.)

Testing for Viruses

Viruses are obligate intracellular parasites; they can grow only inside animal or plant cells. This growth requirement on their part necessitates sophisticated testing methods and specialized laboratories. Testing for human viruses involves adding concentrated samples to small flat-sided flasks prepared with thin monolayers of tissue cells. These cells are often taken from malignant growths of cancer patients and processed for commercial sale. Through microscopic analysis, specially trained personnel look for distinctive types of cellular damage unique to certain viral genera or groups. This damage is called the virus's *cytopathic effect*, or CPE.

Molecular biological techniques are now being combined with viral cultures in an attempt to improve accuracy, but these techniques add significant cost to the tests, and they require skilled scientists. Performing virus studies on water samples is difficult because there are many interfering factors, such as the suspended soil that rivers commonly carry. For this reason, many utilities do not perform this testing, or they defer it to larger laboratories with greater expertise. These laboratories, however, have greatly improved the culturing of viruses in the last few years and have attained an impressive isolation sensitivity, especially when assessing finished water.

In place of performing cultures and other procedures for virus assessment, utilities rely on the proper functioning of their various treatment steps. Membrane filtration systems, ultraviolet light applications, and ozonation are effective in inactivating viruses, but most utilities rely on the final step of chlorination. If a utility's operators are proficient at conducting clearwell chlorination through the proper application of CT values based on water temperature, chlorine concentration, and time, they may be confident that pathogenic viruses and other disease organisms have been inactivated.

The correlation of chlorine contact time values with microbiological conditions is empirically derived. In other words, the CT values are used to indirectly determine the killing or inactivation of harmful microorganisms by relating those CT values with data derived from previously conducted studies by research scientists.

Testing for Coliphages

Some viruses attack only bacteria. They are called *bacteriophages*. A type of bacteriophage that is helpful in assessing the biological nature of water is called a *coliphage*. These coliphages only target *E. coli*. If they are present in high numbers in water, they can be used as surrogate organisms to suggest the presence of *E. coli*.

What is especially useful is that coliphages have the size and physiological characteristics of human viruses. Their presence, therefore, represents the beneficial or hostile nature of the water environment to the coliphages and indirectly to infectious human viruses.

The method for growing coliphages is lengthy. Special strains of *E. coli* are grown and added to melted agar. Next, the sample or control organisms are added to the flask of agar and poured into petri plates. These agar plates are incubated overnight and read the next day.

The *E. coli* that is in the agar grows in a thick, uniform lawn and is available as food for the coliphages. If the coliphages are present, a clear zone called a *plaque* will be present where the *E. coli* has been killed. These plaques are counted to derive a quantitative value for the coliphages present in the sample (Figure 4-19).

The successful attainment of precision with the coliphage test by a scientist can be difficult. A coliphage is tiny, representing only one 800 quadrillionth of a 100-mL sample. This explains why variation of recovery within a narrow range of only a few hundred can be experienced.

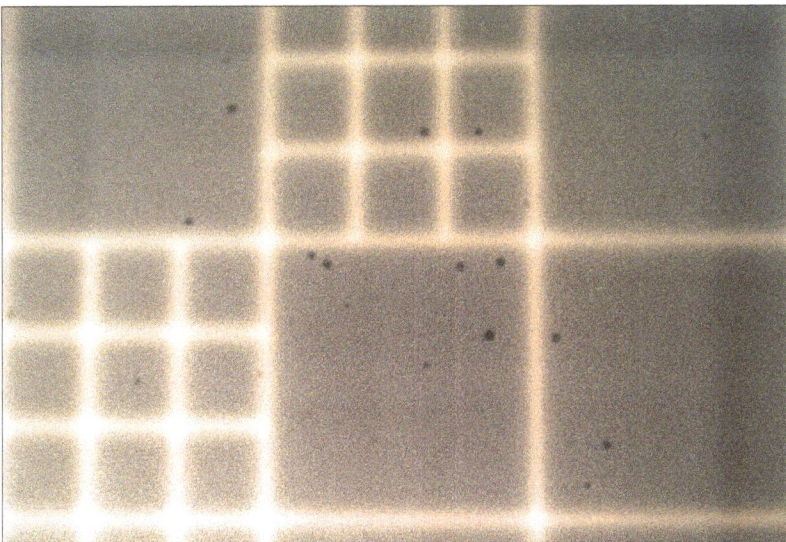

Figure 4-19   Coliphage-MS2 plaques on a lawn of *E. coli*

Methods of Testing for Other Organisms

Water industry personnel are seldom concerned with fungi. These ubiquitous organisms can grow in adequately moist and warm environments with very few nutrients. Most laboratory tests for these organisms attempt to grow the fungi on agar plates and study the resulting growths under the microscope. Because of their increasing clinical importance, molecular biological techniques are being developed to identify fungi.

Advances in Molecular Biology

The science of molecular biology has developed techniques for detecting certain species and groups of microorganisms by identifying their deoxyribonucleic acid (DNA). DNA probes detect specific chromosomal fragments known to be unique to specific organisms. Some of these methods are simple to use, while others are expensive and time-consuming.

These analytical techniques hold great promise for improving detection and identification of particular pathogens. Most laboratories are waiting, however, until more streamlined and less expensive molecular biology methods become available before replacing their conventional techniques of culturing samples in nutrient media. Some companies have developed high-tech instruments; however, the instruments can be used only after laboratory processing or culturing of samples.

Light Microscopy

Light microscopes have long played a powerful role in studies of microbiological processes. Despite continuing innovation in analytical methods, they remain essential tools for discovery and sample analysis.

Biochemical, metabolic, and molecular biological applications can often identify characteristics of organisms far beyond what may be observed with a microscope. Yet sophisticated instruments retain essential advantages as well. Speed and ease of use are strengths of microscopic analysis. This familiar method remains a reliable way to confirm the presence of certain organisms, crystals, and miscellaneous organic particles.

Antonie van Leeuwenhoek was the first to discover microorganisms using unique hand-built microscopes in 1674 (Figure 4-20). Prior to that time, mites were the smallest known creatures. His discovery was the beginning of a slow emergence of the understanding that diseases could be caused by microorganisms.

Leeuwenhoek began thinking about microscopy when he noticed merchants using magnifying glasses to judge the quality of cloth before purchase. This practice aroused his interest and inspired him to build

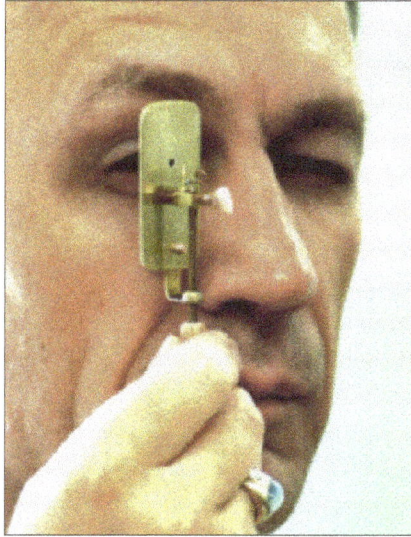

Figure 4-20   Leeuwenhoek microscope

miniature microscopes with tiny, highly refined lenses ground painstakingly by hand. Some lenses were ground with single grains of sand. Leeuwenhoek also extracted metal from ore, then melted and shaped the metal himself. He had produced 400 microscopes by the time he died at age 91.

Using these instruments, he studied a variety of materials from human blood to elephant tusks. His discovery and observations of bacteria and protozoa, using lenses as small as pinheads, proved to be his greatest discoveries. Leeuwenhoek enthusiastically reported his findings throughout his life to the Royal Society of London, but neither he nor others made a connection between microbes and diseases. Illnesses that impaired and killed huge numbers of people were still thought to be caused by a large assortment of evil vapors and unclean thoughts.

Finally in 1876, 200 years after Leeuwenhoek's discovery, Robert Koch and Louis Pasteur made the connection between microorganisms and disease, establishing the concept known as the *germ theory*. This leap in understanding allowed them and others to launch an attack on diseases, which has greatly improved human health and life expectancy.

The design of microscopes progressed throughout the 20th century. Monocular instruments made of brass were popular in the early 1900s, while more advanced binocular instruments were developed in the 1950s. Today, the technology of microscopy continues to advance with the addition of numerous lenses, filters, and fine adjustments (Figure 4-21). Some have evolved into complete computerized workstations.

Figure 4-21   Modern microscope

Particle analysis of water is a great example of the value of microscopy in evaluating the varying and unpredictable characteristics of water samples. It allows the discovery of particles and organisms that often might not be targeted or detectable using chemical analysis or microbiological techniques. Careful counts can yield quantitative results. Few other methods can supply such rapid and practical guidance for water treatment decisions.

Alternative test methods have indirect results compared with microscopy. Other tests may detect signs associated with certain pathogens and then infer the presence of the organisms. As this link becomes more indirect, the occurrence of false-positive and false-negative results increases. Sometimes the presence of a specific pathogen may never be directly confirmed. Even more significantly, individual tests might detect only those organisms specifically targeted and might not reveal all organisms in the sample.

Kohler Illumination

Whether one is using a simple or a complex microscope, the micro-scope must be properly adjusted to maximize image quality. Much of this depends on the proper application of the technique called *Kohler illumination*.

To accomplish this, first focus on a microscope slide with a sample. Close the field iris (lower) diaphragm until the shutters are in the field of vision. Raise or lower the condenser aperture iris diaphragm (upper) assembly using the diaphragm side knob until the shutters are in focus. Reopen the field iris diaphragm until its shutters are no longer visible.

Next, open the aperture iris diaphragm. Begin closing it until the image field just begins to darken.

If the microscope has DIC (differential interference contrast) optics, adjust the DIC knob until the image is satisfactorily lit.

Aerobic Endospore Method

This valuable test can give water utilities an idea of how well their treat-ment processes are performing relative to the empirically derived *Crypto-sporidium* credits the USEPA assigns.

Soil and most natural water bodies commonly contain large numbers of aerobic bacteria that produce endospores. The great majority of endospores are produced by *Bacillus* species, but this method does not differentiate the various species, and thus the generalized method name.

These environmentally resistant endospores can be used as biological surrogates for *Cryptosporidium* oocysts. They each have a similar resis-tance to water treatment. Whereas the originally published endospore method (Rice et al. 1996) required a cumbersome and time-consuming sample-heating step, the following method was modified by the author to replace this step with post-filtration heating of the membrane filters. The result is a test that requires much less time and sample manipulation.

This method also allows the processing of large amounts of water, which significantly increases the sensitivity of the method. This is important because one wants the endospores to persist through the entire process, thus making it possible to calculate the overall removal of endo-spores, expressed as the $\log_{10}$ *Cryptosporidium* removal potential.

Since this method has been streamlined, it is now feasible for labo-ratories to employ it as a routine daily or weekly test, thus frequently monitoring a water treatment plant's treatment effectiveness relative to *Cryptosporidium*. Furthermore, the generated sets of data become valu-able records for assessing the effects of seasonal or treatment chemical changes.

Collect the samples to be studied, such as lake or river water, basin effluent, filter effluent, and finished water. Estimate volumes that might be necessary for each sample. It is best to begin with the lake or river, calculating approximately 8 endospores/ntu for a Midwestern river with a suspension of rich soil. Aim for 30–180 colonies per filter. Experimentation will be necessary to determine the levels of endospores that each utility will normally recover.

Example #1 and Procedure

River turbidity = 30 ntu
30 ntu × 8 = 240/mL
Target of 30 colonies/240 = 0.125 mL
Target of 60 colonies/240 = 0.25 mL
Target of 180 colonies/240 = 0.75 mL

Basin effluent might require the use of 100 mL of sample.

As the water is treated, larger amounts of water should be filtered; until when reaching the filter effluent and finished water samples, adequate endospore recovery might require filtering 1–3 L onto a filter.

Arrange the samples starting with the finished water, and filter the samples using 85-mm membrane filters with a 0.45-μm pore size and a 1-L membrane filtration funnel (Figure 4-22).

Using an alcohol-flamed forceps, set each filter upon a sterile nonstick cookie sheet (Figure 4-23). Rinse the funnel with sterile distilled water between each sample.

Once all filters are in place, insert them into an 80°C oven for exactly 15 min to kill existing vegetative bacterial cells, leaving the endospores. Heating at too high of a temperature or for too long a time will kill the endospores. The filter identification can be written at the edge of the filter with an ink pen prior to use.

Using a sterile forceps, grasp each filter and apply it to plated tryptic soy agar medium with an additional 0.4 percent agar (total of 1.9 percent) to help minimize spreading of colonies. The filters curl or distort upon drying but will lie flat once again when rolled onto the moist medium.

Incubate the agar plates with filters at 35°C. At 22–24 hours, count the colonies using a hand lens (Figure 4-24).

A counting cover can be made by marking an extra petri dish cover into quadrants with a marker. This device helps separate the colonies into areas so that they are better visualized. Wipe the inside of the counting cover with a 10 percent solution of Tween 80 to prevent condensation, which might interfere with visualization of the colonies. Allow the Tween 80 solution film to dry, and the counting cover will perform well for several months (Figure 4-25).

Figure 4-22   Membrane filtration apparatus

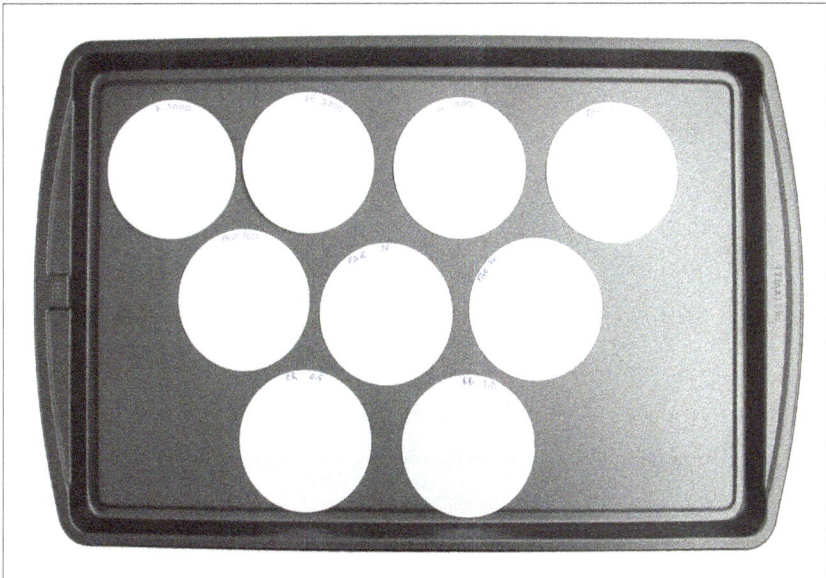

Figure 4-23   Cookie sheet with filters

Figure 4-24   Colonies derived from endospores

Using Excel's $\log_{10}$ formula or an appropriate calculator, calculate the $\log_{10}$ removal from one treatment step to the other by first determining the fold decrease in endospores. Insert this number into the parentheses of the formula and it will calculate the $\log_{10}$ change.

Example #2

River water count: 28,000/L
Treatment basin effluent: 220/L
28,000/220 = 127.3 fold decrease
$\log_{10}(127.3) = 2.1 \log_{10}$ removal

Method-specific supplies:

- 1,000-mL Nalgene sterilizable sample bottles
  (e.g., Fisher Scientific cat. #02-924-206)

- 1-L membrane-filtration funnel (e.g., Fisher Scientific cat. #09-753-2)

- 85-mm Millipore Immobilon membrane filters with a 0.45-μm pore size (Millipore cat. #HATF08550)

- Nonstick cookie sheet(s)

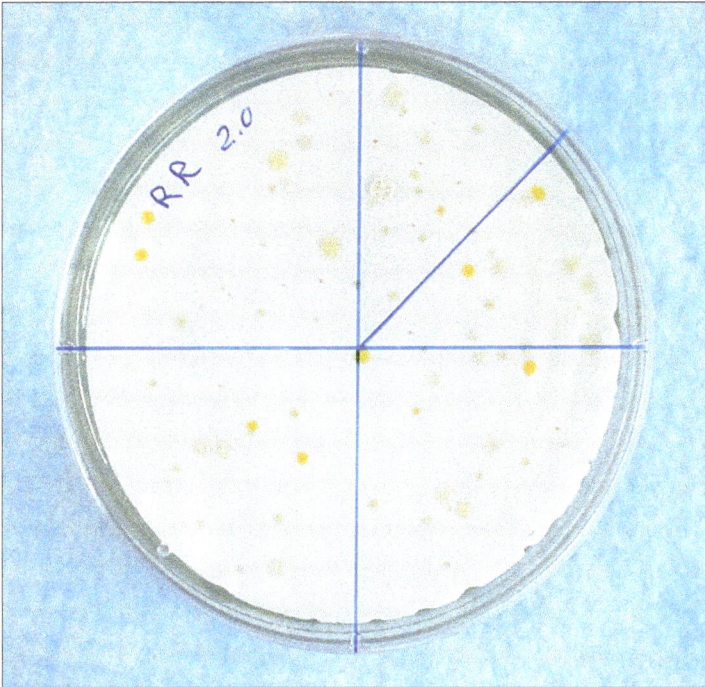

Figure 4-25   Endospore counting cover

- Tryptic soy agar (e.g., Difco cat. #0369-17-6) with an additional 0.4 percent agar (total of 1.9 percent) to help minimize the spreading of colonies

Common laboratory supplies:

- Incubator adjustable to 35°C
- Oven able to accurately maintain 80°C, and large enough to accommodate nonstick cookie sheets
- Pan balance, hot plate, flask, autoclave, and 100 × 15-mm petri dishes, for medium
- Vacuum manifold and vacuum source
- Forceps
- Alcohol for flaming forceps
- Flame source, such as a small hand propane torch or Bunsen burner
- Sterile distilled water to rinse funnel between samples
- Timer for oven
- Hand lens

Well Slide and Glycerol Phytoplankton Method

Accurate plankton studies (especially of algae and cyanobacteria) from natural water are valuable when assessing the water's suitability for treatment,
or for determining possible sources of stream phytoplankton blooms.

Accurate phytoplankton studies from the filter effluent and finished
water of drinking water utilities are valuable when assessing thorough-
ness of treatment or when troubleshooting episodes of taste-and-odor
problems, high filter turbidity, or biofilm development in the system.

Existing methods present various difficulties that are avoided when
using the simple, fast, and accurate well slide with glycerol method presented here.

Traditional methods that employ settling techniques or the study of
slides with relatively large chambers (Sedgwick-Rafter slides) present the
phytoplankton beautifully and with great detail. This feature is especially
useful when extensive speciation is desired by professional phycologists.
The well slide with glycerol method presented here is not for similar professional use but will be very useful for utilities interested in generating
accurate algal and cyanobacterial counts with adequate identifications
within a couple of hours after collection, thus providing relevant results
for real-time operation decisions.

Procedure

Well slide preparation. Prepare well slides that have 5-mm diameter wells
for sample receipt, by adding 10 µl of an acid and glycerol mixture (0.5 mL
glycerol and 20 mL 0.1N HCl). The acid is to reduce crystal formation and
improve cellular morphology. The glycerol provides a thin film in which
the phytoplankton cells remain without drying.

River and lake sample preparation. Gently vortex the sample and
draw off 50 µL using a microtiter pipet. Deposit the water onto a glycerol-
primed well (Figure 4-26). A spherical drop will form. Concentrate the
sample through evaporation, by placing the slide on a shelf in a 30–35°C
incubator, or on a rack slightly elevated above a slide warmer or a coffee
cup warmer providing similar air current and temperature.

The latter techniques are more convenient because they circumvent
having to move the slide while it contains the large water droplets. Heating the sample beyond the recommended temperature might damage cell
morphology and gel the glycerol.

Once the water has evaporated, the microbes remain uniformly distributed in the thin layer of glycerol, which does not dry. Do not cover-slip
the well slides.

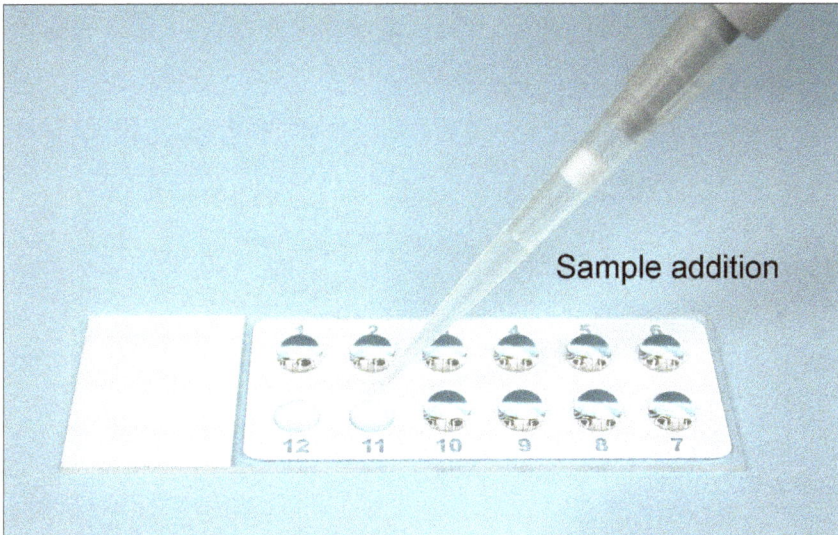

Figure 4-26   Well slide with samples

Filter-applied, filter-effluent, and finished water processing. When the well slide and glycerol method is used for testing clear water with only a few organisms per milliliter, such as the filter-applied, filter-effluent, and finished water of drinking water utilities, centrifugation is recommended.

Collect 10 mL filter effluent or finished water in a narrow-tipped glass centrifuge tube (Figure 4-27). (Most plastic centrifuge tubes are too blunt for accurate decanting).

Centrifuge at 1,000 rcf for 20 min to settle the microbial cells. Using a narrow-tipped pipet, aspirate the water down to 6 mm (¼ in.) from the tip. This will leave approximately 50 µL of concentrated sample.

Using a clean Pasteur pipet, gently mix and draw up the entire concentrated sample and deposit it onto the well slide with the acid and glycerol mixture. Concentrate further through evaporation as described for raw water samples. Do not add a cover slip.

Following use, wash and rinse the centrifuge tubes well. End with a 3–4 mL ethanol vortexed rinse to remove any remaining water.

Microscopic analysis. Microscopic analysis is best performed at 200×, using a 20× objective lens paired with 10× ocular lenses. Preferably, the microscope will be equipped with DIC (differential interference contrast) optics or comparable image enhancing filters. These criteria will allow ample magnification for viewing details, create the best lighting conditions, (which can otherwise vary because of the refractive index of

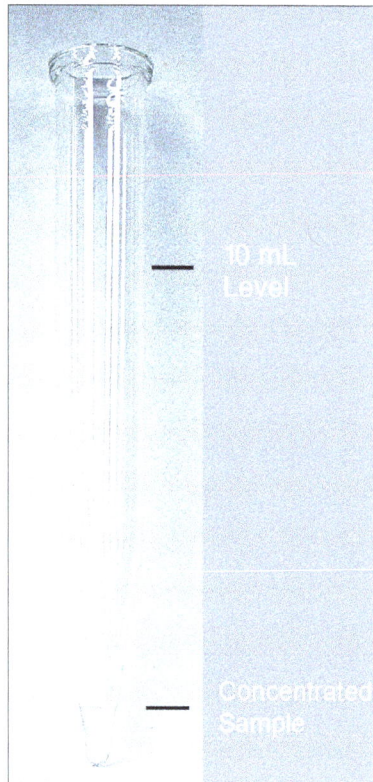

Figure 4-27   Centrifuge tube

the glycerol), and present an ideal field size for counting the microorganisms. The acceptability of alternate microscopic arrangements must be assessed by the scientist making them.

Starting at one side of the slide, systematically scan the slide to view and count the microorganisms. Use the mask (slide paint) as a focusing reference. It will be in the same focal plane as the sample, making it easy to remain focused while studying the slide and thus ensuring no cells are missed due to their presence in a different focal plane. Furthermore, the surface of the slide will be discernible even with relatively debris-free samples. Because each well of the slide is small, assessment of the entire applied sample is easily accomplished. A count of the complete well eliminates inaccuracies relating to cell distribution.

Some microbiologists have found it better to scan slides vertically. Scanning up and down a slide helps eliminate unintentional duplicate counting, and better conforms to natural eye movement.

Count individual cells of groups or filaments, such as the trichomes of cyanobacteria. For example, the cyanobacterium *Planktothrix* often has trichomes that range from 10 to 60 cells. Furthermore, algal colonies will be composed of several cells, which fact warrants the counting of individual cells, such as with *Pediastrum* or *Scenedesmus* (Figure 4-28).

A two-key lab counter may be used so that easily managed separate counts of algae and cyanobacteria may be made.

A complete count of each well with an initial sample volume of 0.05 mL river or lake water, multiplied by 20, will give the total count per milliliter. A complete count of each well with filter effluent or finished water, divided by 10, will give the total count per milliliter.

Aphanizomenon is a cyanobacterium that most particularly forms mats on water surfaces without normally dispersing through the water column. This characteristic can make the proper enumeration of this organism difficult. The sample should be mixed well, and the cells of individual fascicles enumerated. One fascicle ¼ in. long might be 100 cells wide and 800 cells long, equaling 80,000 cells or so. Filtering a measured amount of sample using a membrane filter allows one to estimate the number of fascicles per milliliter.

Method-specific supplies:

- Well slides with 5-mm diameter wells (e.g., Fisher Scientific; 12 well slides: cat. #12-580-23)
- Slide warmer or a coffee cup warmer
- Narrow-tipped glass centrifuge tube (e.g., Fisher Scientific; cat. #05-538-410 or 05-569-3) if intermediately treated or finished water is to be studied
- Two-key lab counter (e.g., Fisher Scientific; cat. #02-670-12)

Common laboratory supplies:

- Microscope with 20× objective lens
- Microtiter pipet able to deliver 50 μL
- Microtiter pipet able to deliver 10 μL
- Centrifuge if intermediately treated or finished water is to be studied
- Glycerol
- Hydrochloric acid
- Pasteur pipets and bulbs

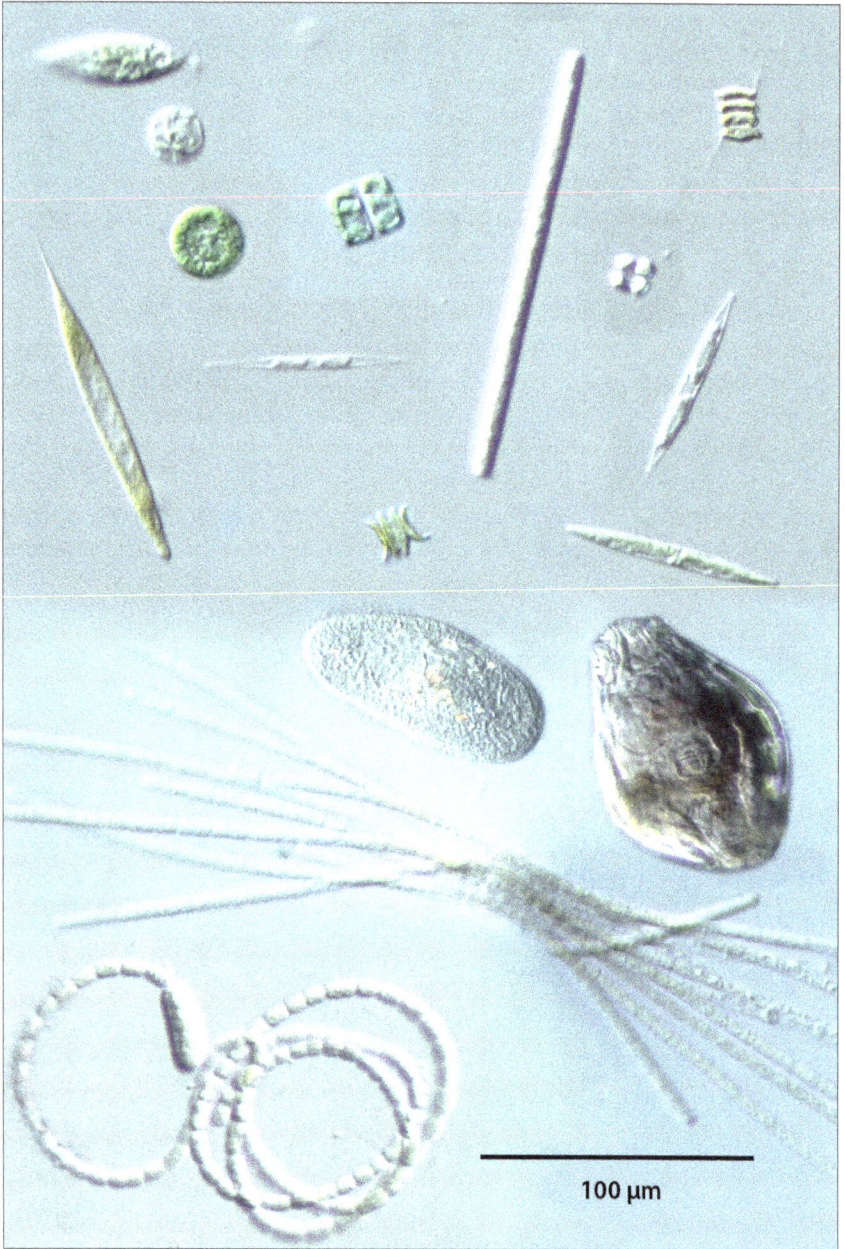

Figure 4-28   Collage of microbes on well slide, 200x

CHLORINE CONTACT TIME VALUE

Accurate chlorine contact time values calculations ensure effective viral removal (enteroviruses, adenoviruses, reoviruses, hepatitis viruses, etc.), as well as effective bacterial, protist (excluding *Cryptosporidium*), and miscellaneous microorganism inactivation.

It is imperative that each utility calculates its daily CT values and monitors the chlorine level of its clearwells continuously.

To do this requires detailed knowledge of the volume and flow of finished water through the utility's clearwell. If these data were not supplied by the engineers who first constructed the clearwell, an operator or manager must determine them and develop charts of target chlorine or alternative disinfectant amounts relative to water temperature and flow rate. Treatment plant operational manuals discuss this in detail.

Continuous daily records of disinfectant applications are necessary to prove proper operational procedures. Failure to diligently engage in the proper application of these chemicals might result in the illness or death of customers who rely on their community's drinking water treatment plant for safe water. There are incidents where negligence at performing proper water disinfection has resulted in operators being imprisoned for their failure to meet their responsibilities.

EFFECTIVE DATA COMMUNICATION

Some utilities are small enough that the operators who conduct laboratory assessments are also those who apply the results to the plant operations. Other utilities have laboratory personnel who generate the data and who must communicate this data to the plant operators.

The laboratory personnel who generate data for other plant employees must at least put this data on records where the data are effectively communicated. It is common for these data to be added to laboratory data files for access by others. In addition to this practice, the microbiologist of Des Moines Water Works chooses key microbiological parameters that are added to an Excel spreadsheet called the *Microbiological Treatment Effectiveness spreadsheet* (Figure 4-29).

Included in the spreadsheet are the river turbidities, the aerobic endospore/*Cryptosporidium* removal determinations for each step of treatment, the phytoplankton (algae and cyanobacteria) enumerations for each river, and each step to treatment, the total coliform and *E. coli* counts for the filter effluent water, chlorine contact time data, and additional comments relating to any one day's source water quality or treatment effectiveness.

Microbiological Treatment Effectiveness

EPA considers microbiological parameters to be the foremost critical for drinking water utilites to control.

EPA scientists have developed a method for using aerobic bacterial endospores as surrogates for *Cryptosporidium* and overall treatment plant performance. They determined through various studies of numerous parameters that the three most valuable indicators of treatment efficiency are aerobic endospores, turbidity, and particle counts. Total coliform bacterial counts and *E. coli* counts are considered the most directly related to bacterial pathogen removal. Phytoplankton counts aid operators in choosing water sources. High counts of either cyanobacteria or algae can rapidly clog filters and shorten their cycle life.

Algal numbers relative to filter problems: Light = 0- 10,000, Moderate = 10,000- 20,000, Heavy = 20,000-30,000, Very Heavy >30,000/ml.
Cyanobacteria: >500 but <2,000= Level I (T&O), >2,000 but <15,000= Level II (Toxin studies), >15,000 per ml= Level III (Contingency plan).

The target values included on the data charts are based on low average values determined by past DMWW plant functions. They are conservative targets to help show when treatment efficiency is waning. Based on several river studies that showed a low incidence of *Cryptosporidium* in our source waters, EPA determined the target removal values do not have to exceed $3Log_{10}$ (1,000 fold), for our three DMWW treatment plants, but this does not mean it is safe to operate at this minimal level. A broken sewer line or discharged agricultural waste lagoon could threaten our plants with pathogens that exceed a $3Log_{10}$ removal potential. Furthermore, excessive phytoplankton will reach the filters at $3Log_{10}$.

Fleur Plant Treatment Effectiveness

	River Turbidity		Aerobic endospore assessments						Phytoplankton				Bacteria		Viruses	Diminished Treatment Notice
	RR Turbidity NTU's	DMR Turbidity NTU's	River plus Gallery INFLUENT $\geq 0.3 \ Log_{10}$	Combined Influent to FA BASINS $\geq 2.3 \ Log_{10}$	FA to FE FILTERS $\geq 0.8 \ Log_{10}$	FE to F CHLORINE $\geq 0.5 \ Log_{10}$	Influent to FE CRYPTO. EQUIVALENT $\geq 3.4 \ Log_{10}$	Influent to F OVERALL & B.ANTHRACIS EQUIVALENT $\geq 3.9 \ Log_{10}$	RR Cyano-bacteria	RR Algae	DMR Cyano-bacteria	DMR Algae	FE Total Coliform $\leq 10/100ml$	FE E. coli $\leq 3/100ml$	Clearwell chlorine Contact Time A=Acceptable	
Targets →																
2/6/2012	4.3	5.2	0.3	2.3	0.8	0.5	3.4	3.9	1000	300	500	200	10	3	3 A	
2/7/2012	9.2	8.1	0.3	2.0	0.8	0.5	3.1	3.6	800	200	600	400	10	1	1 A	Evaluate!
2/8/2012	8.2	10	0.3	2.5	1.0	0.5	3.8	4.3	600	100	800	900	8	2	2 A	
2/9/2012	15.3	10.5	0.4	2.3	0.9	0.6	3.6	4.2	700	100	800	1000	6	2	2 A	
2/10/2012	25.2	20.5	0.4	2.4	1.1	0.8	3.9	4.7	800	200	900	800	7	4	4 A	Evaluate!
2/11/2012	24	21.2	0.3	2.3	1.0	0.5	3.6	4.1	900	500	600	600	8	3	3 Unacceptable!	
2/12/2012																
2/13/2012																

RR—Raccoon River, DMR—Des Moines River, FA—Filter Applied, FE—Filter Effluent, F—Finished Water

Figure 4-29 Spreadsheet of microbiological data
Source: Des Moines Water Works (DMWW), 2012 data

Furthermore, charts of some of this data are often inserted for additional interpretation convenience.

The columns of data for each parameter have conditional formatting, so that bars are shown for each day's data and its relative amount to previous data for that parameter. In addition, when a test value exceeds the maximum desired target listed in the column's heading, the number is shown in red, and the comment *Evaluate!* appears in red in a column to the right.

This spreadsheet makes it convenient for the operators to readily assess the treatment plant's microbiological treatment effectiveness without having to digest the same data that are often randomly listed in a database. These data do not always correlate with simple chemical determinations of sedimentation or coagulation effectiveness using chemical titration methods, and so on.

Since the microbiological purity of drinking water supersedes its softness or taste in importance, wise operators assess the microbiological parameters and adjust the plant chemicals accordingly.

MEASUREMENTS AND DATA ACCURACY

All analytical methods are limited by their inherent relevance and maximum accuracy. These characteristics affect the reliability of testing from the sampling stage to the final result. Data generated from water sample studies must be expressed using these concepts.

Data relevance is an expression of how effectively an operator can use test results. Data should not be reported at accuracy levels more specific than users of the information can practically apply. For example, results from a daily test on a water component might fluctuate several whole units every hour (e.g., rising from 60 to 70 units). The report of that test need not record a result more specific than a whole number (e.g., the report might state 65 rather than 65.32). Even if a laboratory instrument accurately gives the result to the hundredths place, the reading should be rounded to the nearest whole number because the additional precision cannot improve practical control of treatment.

Data accuracy is as important as its relevance to treatment decisions. It depends on specific practices for sample collection and measurement. Testing cannot generate a final result more accurate than the least accurate measurement made during the study. When measuring a sample, the smallest pipet that can hold the entire volume should be used. A test result loses accuracy in proportion to the number of measurements taken of each sample.

When a sample is measured at 100 mL with a graduated cylinder having an accuracy of plus or minus 5 percent, the true volume lies between

95 mL and 105 mL. If the entire sample were cultured for bacteria, a micro-biologist should not be overly concerned whether the colony count is, for example, 42 or 43. The limited accuracy of the initial sample measurement prevents any greater accuracy of results.

When different values are added together, the final result must be reported with the number of decimal places of the least accurate value. For example, with the series 2.30 + 4.322 + 3.0 + 2.1213 = 11.7433, the sum should be reported as 11.7.

When different values are multiplied, the result should be expressed relative to the least significant numbers in the factors. Refer to the chapter on Significant Figures in *Standard Methods for the Examination of Water and Wastewater* (APHA, AWWA, WEF 2012) for a detailed discussion.

Round off numbers with final digits ranging from 1 to 4 downward. Round off numbers with final digits ranging from 5 to 9 upward. For example, for 3.71 through 3.74, round to 3.7 in a report listing values to one decimal place; 3.75 through 3.79, round to 3.8.

If a number is generated with a particular accuracy, any report should record it to the correct number of decimal places, even if the last digit is zero. This practice indicates to anyone reading the report that testing was performed to the indicated accuracy. For example, record 3.20 instead of 3.2 for values carried to two decimal places.

Calculators and computers often allow users to automatically round values to the nearest nonzero number. This adjustment is sometimes undesirable, and settings should be adjusted to give values at the appropriate level of accuracy.

Data entry and reporting should also avoid hanging decimal points. This condition occurs when a decimal number less than 1 is recorded without a zero preceding the decimal point. The absence of a whole number increases the likelihood of data being misread. For example, rather than recording .235, record 0.235.

CHAPTER 5

Chemistry of Microbiology and Water Treatment

The great assortment of elements that comprise the periodic table of the elements, and consequently all living and nonliving substances of the universe, originated in the stars. Most large stars create enough pressure and energy to forge helium, carbon, oxygen, nitrogen, iron, and other elements from the hydrogen that fuels their nuclear fusion. The creation of the rest of the elements, and the distribution of them throughout the cosmos, require supernovae explosions. Planets coalesce from the remnants, and thus all else evolves.

The basic laws of physics define the processes of chemistry. Similarly, chemical processes define the basic functions studied in the life sciences. For this reason, the fields of chemistry and microbiology overlap in critical ways. This relationship creates a need for water facility personnel to learn something about the basic principles of chemistry. It will help them to understand the behavior of pathogenic microbes as well as the numerous disinfection methods. Knowledge of chemistry also helps to explain how the functions of beneficial microorganisms contribute to a healthy ecosystem.

This chapter provides an introduction to the elementary concepts of inorganic and organic chemistry. It also outlines the natural process of nitrogen fixation in the environment. The last sections of the chapter discuss the chemistry of water softening and the disinfectants used by water facilities. For more detailed instruction on these topics, consult the water treatment manuals listed in the bibliography or refer to general chemistry texts.

INORGANIC CHEMISTRY

Inorganic chemistry is based on a basic set of elements. These chemical elements are composed of atoms, the smallest particles that retain an element's original characteristics. Atoms can be broken down further into subatomic particles—the proton, neutron, and electron—but these building blocks of matter are stable only as parts of atoms.

Protons carry positive electrical charges. They combine with neutrons, which have no charge, to form the nucleus of an atom. Negatively charged electrons occupy the space surrounding the nucleus in a series of "shells" that surround the nucleus. If the atom were the size of a football stadium, the nucleus would be no larger than a small insect flying in the middle.

The defining characteristic of an atom of any element is the number of protons in the nucleus, called its *atomic number.* An atom is electrically stable when it has the same number of electrons as protons, so their opposite charges balance one another, and the atom itself has no charge. An electron may be removed from or added to the outer shell of an atom. This type of atom is called an *ion.* An ion with more protons than electrons has a positive charge and is called a *cation.* An ion with more electrons than protons has a negative charge and is called an *anion.* The number of electrons in an atom's outer shell determines how the element combines with others to form compounds.

Periodic Table of the Elements

Chemical elements have been classified in a highly organized chart called the *periodic table of the elements* (Figure 5-1). Elements are arranged numerically in the periodic table from atomic number 1 (hydrogen) to atomic number 103 (lawrencium). Each successive atom in this series is larger and heavier than the preceding one. Abbreviations of one or two letters represent each element's name and other data often are included as well.

In particular, each element's atomic weight is listed. This number represents the average sum of protons and neutrons in the atom's nucleus. The number of protons always remains the same for any element, but the number of neutrons may vary. Atomic weight indicates the most likely composition of the nucleus. This value is useful for comparison with other atomic weights, but it is not a measure of mass.

The elements are grouped in the periodic table relative to their general characteristics and tendencies to react with other elements. Elements in each horizontal row, called a *period,* have the same number of electron shells. Hydrogen and helium, the only members of the first period, have only one electron shell. The elements of the second period—lithium through neon—all have two electron shells, and so on.

Key

Atomic Number — 6
Symbol — C
Common Name — Carbon
Atomic Mass (Weight) — 6.941

Atomic weights conform to the 1961 values of the Commission on Atomic Weights

Period	1 IA	2 IIA	3 IIIB	4 IVB	5 VB	6 VIB	7 VIIB	8	9 VIIIB	10	11 IB	12 IIB	13 IIIA	14 IVA	15 VA	16 VIA	17 VIIA	18 VIIIA
1	1 H Hydrogen 1.00794																	2 He Helium 4.002602
2	3 Li Lithium 6.941	4 Be Beryllium 9.012182											5 B Boron 10.811	6 C Carbon 12.0107	7 N Nitrogen 14.00674	8 O Oxygen 15.9994	9 F Fluorine 18.9984032	10 Ne Neon 20.1797
3	11 Na Sodium 22.998770	12 Mg Magnesium 24.3050											13 Al Aluminum 26.981538	14 Si Silicon 28.0855	15 P Phosphorus 30.973761	16 S Sulfur 32.066	17 Cl Chlorine 35.4527	18 Ar Argon 39.948
4	19 K Potassium 39.0983	20 Ca Calcium 40.078	21 Sc Scandium 44.955910	22 Ti Titanium 47.867	23 V Vanadium 50.9415	24 Cr Chromium 51.9961	25 Mn Manganese 54.938049	26 Fe Iron 55.8457	27 Co Cobalt 58.933200	28 Ni Nickel 58.6934	29 Cu Copper 63.546	30 Zn Zinc 65.39	31 Ga Gallium 69.723	32 Ge Germanium 72.61	33 As Arsenic 74.92160	34 Se Selenium 78.96	35 Br Bromine 79.904	36 Kr Kryphon 83.80
5	37 Rb Rubidium 85.4678	38 Sr Strontium 87.62	39 Y Yttrium 88.90585	40 Zr Zirconium 91.224	41 Nb Niobioum 92.90638	42 Mo Molybdenum 95.94	43 Tc Technetium (98)	44 Ru Ruthenium 101.07	45 Rh Rhodium 102.90550	46 Pd Palladium 106.42	47 Ag Silver 107.8682	48 Cd Cadmium 112.411	49 In Indium 114.818	50 Sn Tin 118.710	51 Sb Antimony 121.760	52 Te Tellurium 127.60	53 I Iodine 126.90447	54 Xe Zenon 131.29
6	55 Cs Cesium 132.90545	56 Ba Barium 137.327	57–71	72 Hf Hafnium 178.49	73 Ta Tantalum 180.9479	74 W Tungsten 183.84	75 Re Rhenium 186.207	76 Os Osmium 190.23	77 Ir Iridium 192.217	78 Pt Platnium 195.078	79 Au Gold 196.96655	80 Hg Mercury 200.59	81 Ti Thallium 204.3833	82 Pb Lead 207.2	83 Bi Bismuth 208.98038	84 Po Polonium (209)	85 At Astatine (210)	86 Rn Radon (222)
7	87 Fr Francium (223)	88 Ra Radium (226)	89–103	104 Rf Rutherfordium (261)	105 Db Dubnium (262)	106 Sg Francium (263)	107 Bh Bohrium (262)	108 Hs Hassium (265)	109 Mt Meitnerium (266)	110 Uun Ununnilium (269)	111 Uuu Unununium (272)	112 Uub Ununbium (277)	113 Uut	114 Uuq Ununquadium (289)	115 Uub	116 Uuh Ununhexium (289)	117 Uus	118 Uuo Ununoctium (293)

Atomic masses in parentheses are those of the most stable or common isotope.

Note: The subgroup numbers 1–18 were adopted in 1984 by the International Union of Pure and Applied Chemistry. The names of elements 110–118 are the Latin equivalents of those numbers.

57 La Lanthanum 138.9055	58 Ce Cerium 140.116	59 Pr Praseodymium 140.90765	60 Nd Neohymium 114.24	61 Pm Promethium (145)	62 Sm Samarium 150.36	63 Eu Europium 151.964	64 Gd Gadolinium 157.25	65 Tb Terbium 158.92534	66 Dy Dysprosium 162.50	67 Ho Holmium 164.93032	68 Er Erbium 167.26	69 Tm Thulium 168.93421	70 Yb Ytterbium 173.04	71 Lu Lutetium 174.967
89 Ac Actinium (227)	90 Th Thorium 232.0381	91 Pa Protactinium 231.03588	92 U Uranium 238.0289	93 Np Neptunium (237)	94 Pu Plutonium (244)	95 Am Americium (243)	96 Cm Curium (247)	97 Bk Berkelium (247)	98 Cf Californium (251)	99 Es Einsteinium (262)	100 Fm Fermium (257)	101 Md Mendelevium (258)	102 No Nobelium (259)	103 Lr Lawrencium (262)

Figure 5-1 The periodic table of the elements

The vertical columns in the periodic table define the groups that share similar chemical properties. The groups are noble gases, metals, transitional metals, nonmetals, alkali metals, alkali earths, rare earths, and halogens. For example, chlorine, bromine, and iodine, all of which can be used as disinfectants in water treatment, are members of the halogen group. The elements of this group carry a negative charge and react well with positively charged alkali metal and alkali earth elements, such as sodium, potassium, magnesium, and calcium. These reactions are called *ionic bonding*, and they result in common compounds such as sodium chloride, magnesium chloride, and calcium bromide.

Chemical Compounds

A group of chemically bonded atoms forms a particle called a *molecule*. Atoms of two or more different elements can combine in definite proportions to form a huge range of compounds. In nature, atoms tend to bond together and form a molecule whenever the resulting compound is more stable than the individual elements.

The electrons in an atom's outer shell are called *valence electrons*. They strongly influence its tendency to combine with other atoms. Based on experience, chemists have assigned to every element in the periodic table one or more numbers indicating its ability to react with other elements. These numbers are determined by the number of valence electrons, so they are called the element's *valences*.

As elements combine to form compounds, valence electrons are transferred from the outer shell of one atom to that of another, or they are shared by the outer shells of the combining atoms. This rearrangement of electrons produces chemical bonds. When electrons are transferred, the process is called *ionic bonding*. If electrons are shared, it is called *covalent bonding*.

Valence indicates the actual number of electrons that an atom gains, loses, or shares in bonding with one or more other atoms. For example, if an atom loses one electron in a reaction forming an ionic bond, then it has a valence of +1; if an atom must receive one electron to form an ionic bond, then it is said to have a valence of –1. This number varies for different atoms depending on several factors, such as the conditions under which the reaction occurs and the other elements involved.

Electrons are not lost or gained while forming a covalent bond, so the valence of an atom involved in this type of bond is expressed without a plus or minus sign. Table 5-1 lists common valences for a number of elements.

Table 5-1   Valences of various elements

Aluminum (Al)	+3	Lead (Pb)	+2, +4
Arsenic (As)	+3, +5	Magnesium (Mg)	+2
Barium (Ba)	+2	Manganese (Mn)	+2, +4
Boron (B)	+3	Mercury (Hg)	+1, +2
Bromine (Br)	−1	Nitrogen (N)	+3, −3, +5
Cadmium (Cd)	+2	Oxygen (O)	−2
Calcium (Ca)	+2	Phosphorus (P)	−3
Carbon (C)	+4, −4	Potassium (K)	+1
Chlorine (Cl)	−1	Radium (Ra)	+2
Chromium (Cr)	+3, +6	Selenium (Se)	−2, +4
Copper (Cu)	+1, +2	Silicon (Si)	+4
Fluorine (F)	−1	Silver (Ag)	+1
Hydrogen (H)	+1	Sodium (Na)	+1
Iodine (I)	−1	Strontium (Sr)	+2
Iron (Fe)	+2, +3	Sulfur (S)	−2, +4, +6

Chemical Formulas and Notation

The simplest molecules combine only one type of atom, as when two atoms of oxygen combine to form O_2. The subscript indicates the number of atoms of a particular element. Similarly, two atoms of hydrogen combine with one atom of oxygen to form the compound H_2O. The notation "H_2O" is called a *chemical formula* for water.

Formulas are combined with certain symbols to form chemical equations that describe chemical reactions. When the solid sodium chloride (table salt) dissolves in water, the compound breaks apart into ions and the formula for this reaction is

$$NaCl + H_2O \rightarrow Na^+ + Cl^- + H^+ + OH^-$$

This type of an equation always balances. That is, the left and right sides contain an equal number of like atoms. This reflects the principle that matter is neither created nor destroyed, so the number of atoms of each element going into the reaction must equal the number coming out.

Often, chemical equations are simplified by showing the complete compound formulas on each side instead of showing specific ions:

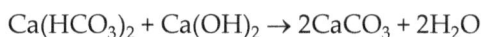

$$Ca(HCO_3)_2 + Ca(OH)_2 \rightarrow 2CaCO_3 + 2H_2O$$

This equation introduces some new notation and a new concept. The parentheses enclose chemical formulas for two radicals. A *radical* is a group of atoms bonded together into a unit that reacts like a single atom with other atoms. The subscript numbers outside the parentheses indicate the number of radicals involved in the reaction.

Also, the number 2 in front of $CaCO_3$ and H_2O, called a *coefficient*, indicates the number of molecules of each compound involved in the reaction. If no coefficient is shown, only one molecule of the compound is involved. Therefore, this equation indicates that one molecule of calcium bicarbonate reacts with one molecule of calcium hydroxide to form two molecules of calcium carbonate and two molecules of water.

This information can be used to add the elements on each side of the equation to confirm that the equation is balanced. On the left, the first compound, calcium bicarbonate $(Ca(HCO_3)_2)$ is made up of one atom of calcium, two of hydrogen, two of carbon, and six of oxygen. Similarly, calcium hydroxide (or lime, $Ca(OH)_2$) includes one atom of calcium and two hydroxyl ions (OH). Each hydroxyl ion includes one atom each of oxygen and hydrogen, for a total of one atom of calcium and two atoms each of oxygen and hydrogen. The grand total on the left side of the equation is two calcium atoms, eight oxygen, four hydrogen, and two carbon.

Carbon atoms have a special ability to form covalent bonds with other carbon atoms and with many other elements. This can create long chain molecules leading to the production of the myriad compounds that are the substance of life.

Acids and Bases

Some substances release hydrogen ions (H^+) when they are mixed with water. This creates an acid, and the reaction produces an acidic solution. For example, when sulfuric acid (H_2SO_4) is mixed with water, many of the molecules dissociate (come apart), forming H^+ and SO^{4-} ions. This release affects the characteristics of the solution, giving it a tendency to oxidize.

Other substances produce hydroxyl ions (OH^-) when they dissociate in water. These substances are called *bases,* and the reaction produces an alkaline solution. Examples that are familiar to water treatment personnel are lime $(Ca(OH)_2)$, caustic soda (sodium hydroxide, NaOH), and household ammonia (NH_4OH).

The acidic or alkaline character of a solution is measured on the pH scale, shown in Figure 5-2. Measurement at the low end of the pH scale represents an increasingly acidic solution; measurement at the high end represents an increasingly alkaline solution. A pH value of 7 indicates a neutral solution, neither acidic nor alkaline. That is the pH level of pure water.

High Concentration of H⁺ Ions	H⁺ and OH⁻ Ions in Balance	High Concentration of OH⁻ Ions

0 — 1 — 2 — 3 — 4 — 5 — 6 — 7 — 8 — 9 — 10 — 11 — 12 — 13 — 14

Pure Acid	Neutral	Pure Base

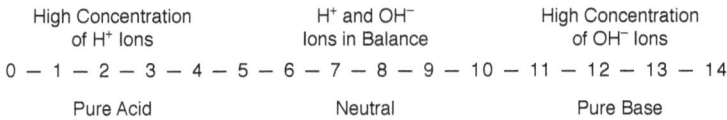

Figure 5-2   Acids and bases on the pH scale

Acids and bases neutralize one another, so pH can be adjusted upward by adding a base or downward by adding an acid. This capability is important for the effectiveness of many water treatment processes. For example, water's pH rises substantially during softening, so it must be reduced again by adding an acidic substance.

ORGANIC CHEMISTRY

Compounds that are produced by life processes are called *organic*. During the previous century, chemists synthesized other compounds that are similar to the molecular structure of carbon atoms and are considered to be organic in nature. This list includes plastics, petroleum products, drugs, and textiles.

Many elements and compounds attach to the carbon atom backbones of organic molecules (Figure 5-3). The vast diversity of these structures has produced a special nomenclature. The prefix in a substance name indicates the number of carbon atoms in the molecule's chain. A name beginning with *meth-* refers to a chemical with one carbon atom. Other prefixes are *eth-* (two carbons), *prop-* (three), *but-* (four), *pent-* (five), *hex-* (six), *hept-* (seven), *oct-* (eight), *non-* (nine), *dec-* (ten), and so forth. In addition, names of some organic chemicals include suffixes that reveal the number of bonds between carbon atoms: *-ane* (one bond), *-ene* (two), *-yne* (three).

The names of some major chemical groups denote similar compounds. Often these compounds attach to longer organic carbon molecules. The list includes alkanes, alkenes, alkynes, aldehydes, ketones, amines, amides, and alcohols.

Suffixes sometimes denote the presence of compounds in these groups. The suffix *-ol* denotes the presence of a compound from the alcohol group, each of which includes one hydroxyl radical (chemical formula: OH). Other suffixes include *-one* for ketones, *-al* for aldehydes, *-amine* for amines, *-amide* for amides, and so on.

The Nitrogen Cycle

Many organic chemicals follow cycles throughout nature. For example, they are assimilated by microorganisms, plants, and animals through metabolic processes and are returned to the soil, water, or air. One of the most important metabolic cycles is the nitrogen cycle (Figure 5-4), which is dependent on microbial activity. Nitrogen compounds support plant growth and, therefore, animal life. They are preserved in the environment through living plant and animal matter to dead organic matter in various stages of decomposition, once again nurturing plant growth. During this process, some strains of bacteria found in soils convert naturally available ammonia (NH_3) to nitrite (NO_2^-) and then to nitrate (NO_3^-).

Because nitrogen is important to plant growth, many commercial fertilizers contain sodium nitrate ($NaNO_3$) and potassium nitrate (KNO_3). Storm runoff carries both naturally produced and artificially introduced nitrates and nitrites into rivers and lakes used as public water supplies. Both compounds have implications for public health, and they are regulated by the USEPA. For example, consumption of nitrite compounds by infants may induce methemoglobinuria (blue-baby syndrome), which dangerously impairs the oxygen-carrying hemoglobin of their blood. Breast-fed babies do not have enough acid in their stomachs to inhibit the growth of bacteria that reduce nitrate to nitrite. Once babies begin eating solid food, they develop levels of stomach acid that inhibit bacteria growth, and they lose their susceptibility to the disease.

In addition, excessive nitrate and nitrite levels may cause large-scale environmental problems. The Mississippi River has high levels of nitrates. During spring runoff, the influx of these nutrients into the Gulf of Mexico causes the growth of great blooms of algae, bacteria, and other microorganisms. The metabolic processes of these organisms deplete the oxygen in the water causing hypoxia, which kills fish and other aquatic animals and significantly impacts the fishing and tourism industries.

To maintain compliance with regulations and ensure supply of a healthy product, water treatment plants must monitor the levels of nitrate and nitrite compounds in their finished water.

Utilities that encounter problems with high nitrate concentrations in their source and finished waters must give ongoing public notices when they exceed 10 mg/L nitrate, or choose a method to reduce the concentration of those compounds.

One method of finished-water nitrate removal employs large vats of resin beads through which the rechlorination water is run. Nitrate compounds adhere to the beads, which are later recharged. This process requires a relatively high capital investment and entails high operational costs.

```
        H                    H  H                  H  H  H
    H - C - H            H - C - H - H          H - C - C - C - H
        H                    H  H                  H  H  H
      Methane               Ethane                 Propane

        H                    H  H                  H  H  H
    H - C - OH          H - C - C - OH         H - C - C - C - OH
        H                    H  H                  H  H  H
      Methanol              Ethanol                Propanol
```

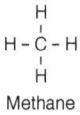

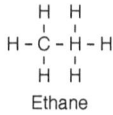

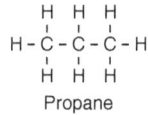

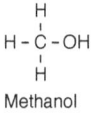

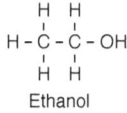

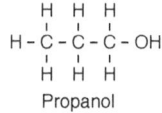

C = carbon
H = hydrogen
OH = hydroxyl radical

Figure 5-3   Structures and names of sample organic molecules

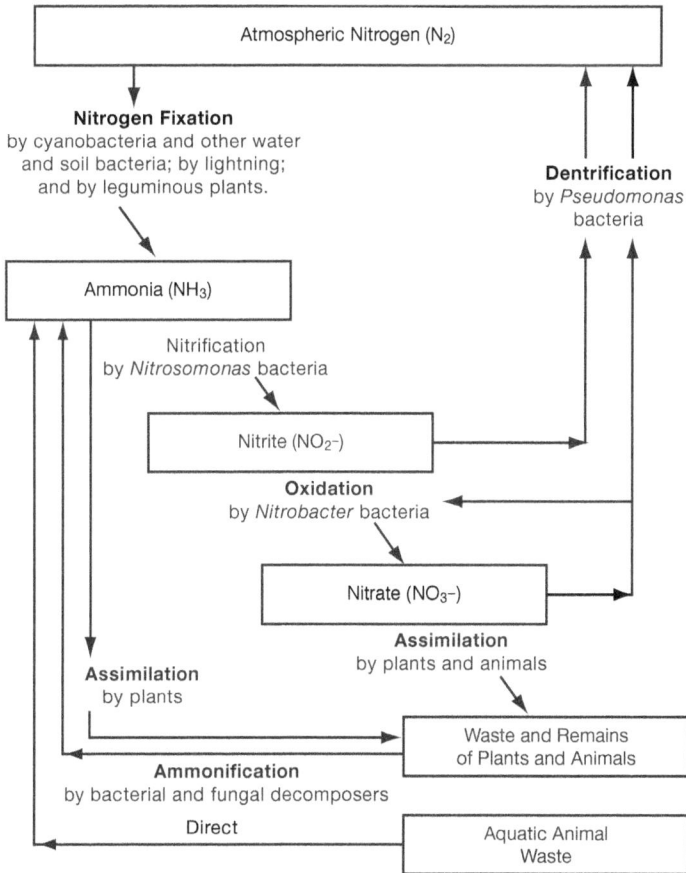

Figure 5-4   Nitrogen cycle

An alternative means of nitrate removal employs the natural biological processes of ponds or lakes. The nitrate compounds are used by microorganisms as nutrients, thus their concentration is reduced.

Water is channeled or pumped into the pond or lake, where it ideally has several days' residence before it is withdrawn for use at the opposite end.

A drawback of this method would be the undesirable growth of cyanobacteria in the body of water being used. To prevent this, one or more water-circulating Solar Bees may be placed in the water. These solar-powered units draw water up a central column and channel its flow horizontally across the pond surface. Cyanobacteria grow best in calm water; therefore, water circulation by the Solar Bees disrupts their bloom potential. Of course, the minimization of nutrient input is also important.

MICROBIOLOGY IN WATER TREATMENT

Why might a utility choose to employ the multiple-barrier approach to water treatment? Why not just concentrate on maximizing one approach?

Microorganisms vary greatly in their sizes, shapes, surface charges, surface textures, motilities, and chemical sensitivities. One treatment method is seldom sufficient for surface-water utilities. It is common for one or two species, out of the hundreds that might be present, to evade initial sedimentation or coagulation steps. These few (especially algae and cyanobacteria) can continue to the filtration process where they might cause filtration plugging, or they might also evade filtration and enter the clearwell. Efforts are made to eliminate them by chlorination before they enter the distribution system.

If inadequate treatment barriers precede clearwell chlorination, the removal of *Cryptosporidium* oocysts will be ineffective, since these hardy oocysts are resistant to chlorine as are other less dangerous microbes.

Surface-water utilities often experience dramatic changes in their source-water rain or snowmelt. This usually requires multiple barriers of treatment to deal with the water quality variety.

Utilities that use groundwater seldom experience these changes and therefore have fewer steps of treatment. If only filtration and chlorination, or only chlorination, are employed, breaches of well casings or the contamination of protected lakes can allow microorganisms to readily pass into the distribution system. In fact, since surface-water utilities have become more vigilant with their treatment for *Cryptosporidium* and other pathogens, groundwater utilities have experienced a greater number of *Cryptosporidium* incidents.

Removing or inactivating microorganisms is an important goal of almost any water treatment process. The addition of lime ($Ca(OH)_2$)

removes suspended solids and dissolved chemicals that cause hardness. This process of softening the water also helps kill or inactivate pathogenic microorganisms. Chlorine or other disinfectants are added to filtered water specifically to prevent the spread of pathogens. Some organisms, notably the parasites *Giardia lamblia* and *Cryptosporidium parvum*, are resistant to disinfection, but effective coagulation and filtration can help to remove them, as can membrane filtration.

Softening

Water treatment often involves application of chemical principles to remove calcium bicarbonate, the cause of carbonate hardness. This is accomplished by adding slaked lime to the water:

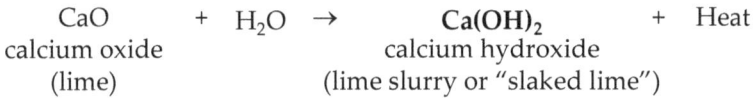

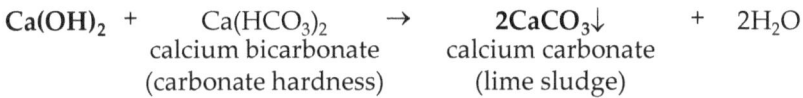

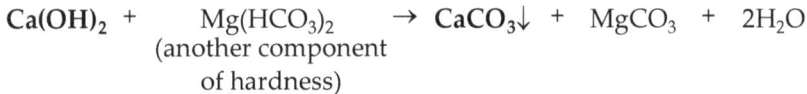

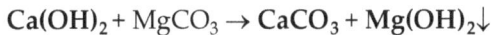

$$CaO \quad + \quad H_2O \quad \rightarrow \quad Ca(OH)_2 \quad + \quad Heat$$
calcium oxide   calcium hydroxide
(lime)   (lime slurry or "slaked lime")

$$Ca(OH)_2 \quad + \quad Ca(HCO_3)_2 \quad \rightarrow \quad 2CaCO_3\downarrow \quad + \quad 2H_2O$$
              calcium bicarbonate   calcium carbonate
              (carbonate hardness)   (lime sludge)

$$Ca(OH)_2 \quad + \quad Mg(HCO_3)_2 \quad \rightarrow CaCO_3\downarrow + MgCO_3 + 2H_2O$$
              (another component
              of hardness)

$$Ca(OH)_2 + MgCO_3 \rightarrow CaCO_3 + Mg(OH)_2\downarrow$$

Note that the downward-pointing arrows ($\downarrow$) indicate that the compounds are insoluble and precipitate out of the water.

Addition of slaked lime also aids in the removal of noncarbonate hardness.

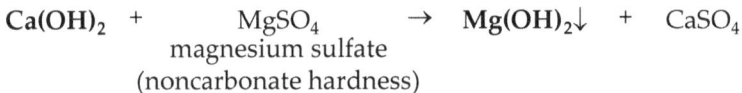

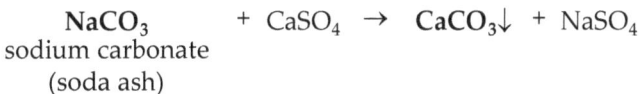

$$Ca(OH)_2 \quad + \quad MgSO_4 \quad \rightarrow \quad Mg(OH)_2\downarrow \quad + \quad CaSO_4$$
              magnesium sulfate
              (noncarbonate hardness)

$$NaCO_3 \quad + \quad CaSO_4 \quad \rightarrow \quad CaCO_3\downarrow \quad + \quad NaSO_4$$
       sodium carbonate
       (soda ash)

These chemical reactions raise the water's pH to between 10 and 11.3. The vast majority of viruses, bacteria, protozoa, and multicellular organisms cannot survive above pH 10.5, so water softening is a powerful disinfectant. Carbon dioxide (CO_2) is added after softening to reduce the water's pH.

Disinfection

Water treatment utilities use one of a variety of disinfectants to ensure pathogenic microorganisms are killed in their finished water. Removal of microorganisms actually begins with coagulation and flocculation. As these processes create floc of suspended solids, some microorganisms are trapped with those particles and removed when the solids settle out during clarification.

More microorganisms are removed when clarified water passes through granular filters. In rapid sand filters, the first few inches of the granular media trap remaining flocs, algae, and other materials, which in turn capture some microorganisms. The schmutzdecke performs the same function in a slow sand filter.

This physical removal process is very efficient in purifying the water, especially if treatment includes softening. In fact, it may be the most effective way within a conventional treatment plant to remove some pathogens otherwise resistant to disinfection chemicals.

Following filtration, the final stage of treatment usually includes disinfection using chlorine, chlorine dioxide, chloramines, or ozone. Some water treatment plants employ new disinfection processes using ultraviolet light (UV) or membrane filtration. Choices among disinfectant practices depend on several factors such as cost, safety, convenience, effectiveness, and the condition of the water being treated.

Chlorine (Cl_2) is the most widely used disinfectant in water treatment. It is relatively cheap, easy to use, and effective at killing most microorganisms present in the water. In addition, some of the chemical remains in the finished water as a free-chlorine residual that prevents bacterial regrowth in the distribution system.

When chlorine is added to pure water, it reacts as follows:

$$Cl_2 + H_2O \rightarrow HOCl + HCl$$
(chlorine) (water)   (hypochlorous acid) (hydrochloric acid)

The chlorine combines with the water to produce hypochlorous acid (HOCl), a weak acid that easily penetrates into and kills bacteria. This action makes chlorine an effective disinfectant. HOCl is also one of two chlorine compounds that act as free available chlorine residual. However, some of the HOCl dissociates as follows:

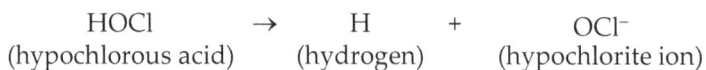

$$HOCl \rightarrow H + OCl^-$$
(hypochlorous acid) (hydrogen) (hypochlorite ion)

The hydrogen produced in this reaction neutralizes alkalinity and lowers pH, while the hypochlorite ion (OCl⁻) is a second type of free available chlorine residual. Its disinfectant action is not as effective as that of HOCl, but it does help to kill microorganisms.

A number of conditions influence the effectiveness of disinfection using chlorine, including pH, water temperature, and contact time. The water's pH, when Cl_2 is introduced, strongly influences the ratio of HOCl to OCl⁻. Low pH values favor formation of HOCl, the more effective free residual, while high pH favors formation of OCl⁻. Similarly, low water temperature slightly favors formation of HOCl.

Contact time is important, along with the concentration of chlorine in the water, in determining the effectiveness of these compounds. Either a longer time or a rising concentration increases disinfectant effect.

The chemical reactions that have been described occur when chlorine is added to pure water. Water being treated with disinfectant also contains other substances, however. For example, Cl_2 reacts with ammonia present in the water to form chloramine compounds.

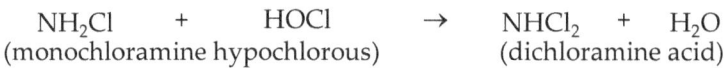

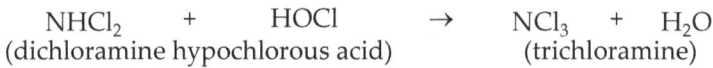

$$NH_3 \quad + \quad HOCl \quad \rightarrow \quad NH_2Cl \quad + \quad H_2O$$
(ammonia hypochlorous acid)   (monochloramine)

$$NH_2Cl \quad + \quad HOCl \quad \rightarrow \quad NHCl_2 \quad + \quad H_2O$$
(monochloramine hypochlorous)   (dichloramine acid)

$$NHCl_2 \quad + \quad HOCl \quad \rightarrow \quad NCl_3 \quad + \quad H_2O$$
(dichloramine hypochlorous acid)   (trichloramine)

The water's pH and the amount of ammonia present determine whether one chloramine compound or more than one are formed.

Monochloramine and dichloramine act as disinfectants, although they are not as effective as free available chlorine, such as HOCl. If contact time is sufficient, chloramines can do an acceptable job of disinfection. However, dichloramine and trichloramine compounds may produce taste-and-odor problems.

Disinfection with chlorine requires additional planning for water with high levels of natural organic material (NOM). Chlorine reacts with NOM to form undesirable disinfection by-products (DBPs). Some of these compounds, especially trihalomethanes (THMs), are regulated as potential carcinogens by USEPA under the Disinfectants and Disinfection By-Products Rule. To avoid creating these compounds, a utility must minimize residual organic material in the water before disinfection and carefully control chlorine usage. Effective coagulation, flocculation, sedimentation, and filtration remove many organic compounds, so chlorine will not react with the compounds to generate THMs. Chloramine reactions do not produce THMs.

Chlorine dioxide (ClO_2) is another very effective disinfectant, and its reactions produce fewer DBPs than those of chlorine. Chlorine dioxide is also more costly and more hazardous to use, however.

Ozone has become a popular disinfectant, despite its high cost, with utilities that have problems with protozoa. In particular, ozone treatment is one of the few effective methods for inactivating *Cryptosporidium*. After ozonation, however, the water contains no residual disinfectant to protect against regrowth of pathogens in the distribution system. Also, ozone cannot be stored, so it must be generated on-site as needed.

Ultraviolet light is another disinfection technology used by some utilities. Under favorable conditions, it inactivates almost all microorganisms without producing undesirable THMs or other DBPs. Once practical only for small facilities, the technology is now being considered at large water utilities. The primary drawback to this treatment is the potential for the UV light bulbs to become coated with light-obscuring material, which prevents the UV light from reaching and killing the organisms. Careful maintenance is needed to ensure efficient operation. Also, turbidity in the water can shield organisms from the UV light, so this technology is practical only in very clear waters. Finally, UV treatment does not leave any disinfectant residual.

Membrane filtration is an effective technology for physical removal of microorganisms from water. After water passes through granular media filters, it is fed through membranes with pores of specific size. Membrane-based water treatment processes (listed in order from larger to smaller pore sizes) include microfiltration, ultrafiltration, nanofiltration, and reverse osmosis. Each removes a progressively finer particle. Ultrafiltration is usually sufficient to remove all microorganisms. No disinfectant residual remains in the water after membrane treatment.

Scientific Nomenclature, Notation, and Numerical Prefixes and Abbreviations

SCIENTIFIC NOMENCLATURE

Scientific nomenclature is arranged from the most general to the most specific. For example:

> Family: Enterobacteriaceae
> Genus: *Escherichia*
> Species: *coli*
> Strain: O157:H7

Use the genus and species to routinely denote microorganisms. The first letter of the genus is always capitalized, and the species name is always in lower case: *Escherichia coli*. The names should also be italicized.

Group names, such as bacteria or enteroviruses, are not considered scientific names and begin with lowercase letters.

SCIENTIFIC NOTATION

Scientific notation is a convenient way to express large numbers by writing them in powers of ten:

> $1 \times 10^1 = 10$
> $1 \times 10^2 = 100$
> $1 \times 10^3 = 1,000$
> $1 \times 10^4 = 10,000$
> $1 \times 10^5 = 100,000$
> $1 \times 10^6 = 1,000,000$

Within this system, for example, 1,200,000 is expressed as 1.2×10^6. The superscript "6" indicates that the decimal point must be moved to the right six places. Similarly, 0.004 is expressed as 4.0×10^{-3}. The superscript "-3" indicates that the decimal point must be moved to the left three places.

NUMERICAL PREFIXES AND ABBREVIATIONS

The following chart lists prefixes and abbreviations for some numbers common in microbiology.

Prefix	Abbreviation	Number
mega-	M	One million
kilo-	k	One thousand
deci-	d	One tenth
centi-	c	One hundredth
milli-	m	One thousandth
micro-	μ	One millionth
nano-	n	One billionth
pico-	p	One trillionth

Parts per million (ppm) = milligrams per liter (mg/L) = 1 gram per million milliliters

Preparation of New
Water Mains for Service

After a new main has been installed according to proper engineering criteria, there still remains the task of cleaning the main of soil and bacterial contamination before it is placed into service. This final operation is crucial for reasons of aesthetics and health.

Customers might tolerate a burst of turbid water flowing from their taps, knowing that a main has been newly laid and that the water will clear, but they will not be tolerant of an illness contracted from that main's water. The first step in providing a clean main is to lay a clean pipe. Next steps are flushing, then uniform chlorination of the new pipe. The final step, having the proper bacteriological tests performed, is the vitally important culmination of the project.

Contractors who experience failed bacteriological tests of new mains often feel frustrated. They do not understand why the tests did not pass after they had "really burned" the pipe with an extra dose of chlorine. This scenario illustrates that chlorination is a complex science, and a laboratory worker needs considerable skill in communicating the proper techniques of new main decontamination and explaining the science behind the operation.

This summary can assist in the understanding of new main decontamination and in the communication between those performing the laboratory tests and those installing the main.

AWWA C651, Standard for Disinfecting Water Mains, provides guidelines for the decontamination of newly installed water pipes. Detailed questions relating to such decontamination procedures should be answered through the reading of the AWWA material.

SOURCES OF CONTAMINATION

Bacterial contamination of water pipes can occur in different ways. Rodents may defecate in the pipes as they lie in storage, soil may enter the pipes as they are installed, and workers may handle the inside of the pipes as they install them. It is also important to use proper joint lubricant. Grease made from animal fat supports bacterial growth and is resistant to flushing because it resists water.

Pipe plugs should be used when mains are being installed. They should be used during lunch-breaks and overnight when pipe ends would otherwise be exposed.

A new main that has been laid for service must undergo several treatment steps.

Debris Flush

A preliminary flush should always be performed. This will remove debris and dirt film from the pipe. Flush until the water has been turned over at least twice. Throughout all flushing procedures, it should be understood that a rapid flow of water through the pipe is crucial. Use the largest "blow-off" nozzle as practical in an attempt to attain a 2.5 to 3 ft/sec flow rate (Figures B-1 and B-2 and Table B-1).

Bullet-shaped foam "pigs" may be used when the pipe is being laid. Foam pigs are forced through the pipe with water, effectively scouring the pipe wall.

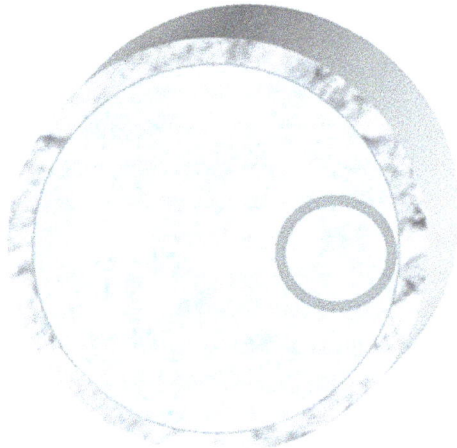

Figure B-1   A 2-in. flush nozzle will move only 1/16 of an 8-in. main's capacity

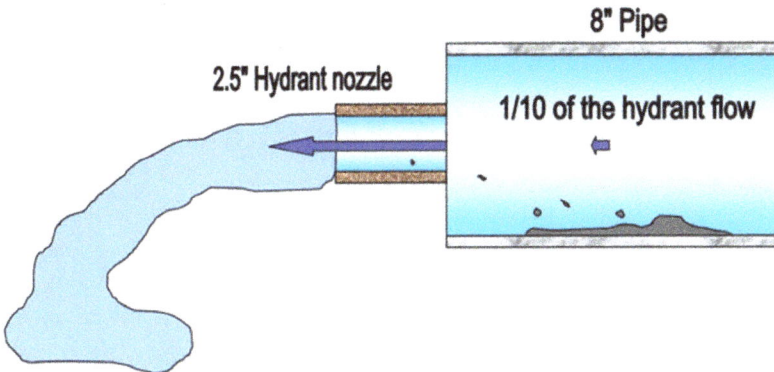

Figure B-2   Main flushing

Table B-1   Flow in gallons per minute from a 2.5-in. diameter nozzle to accomplish 3 ft/sec for the designated pipe diameter

Pipe Diameter (in.)	Flow (gpm)
2	30
4	120
6	270
8	470
12	1,060
16	1,880

Chlorination

Once the initial flush has been completed, it is time to chlorinate the main. A target of 25 mg/L chlorine should be calculated. It is most effective either to inject liquid chlorine using special injection equipment or to pre-dissolve granular chlorine in a barrel, which is then injected into the main.

Add the dissolved chlorine to the beginning of the main and run water until the chlorine is detectable at 25 mg/L at the end of the main. It is important that the chlorine is uniformly distributed throughout the pipe. Seal the main and let the chlorinated water set for 24 hours.

- Chlorine tablets that are glued at intervals throughout a main as it is being laid often fail to adequately dissolve or to properly disperse sufficient chlorine for decontamination. Granular chlorine that has not been predissolved also disperses poorly.

- The use of excess chlorine does not increase the decontamination potential and it risks damaging the gaskets and surfaces of some pipes. Also, once the chlorinated water is flushed from the pipe, it presents an environmental risk. *The contractor is responsible* for studying the area toward which the chlorinated flush-water will be directed. Unacceptable damage to the ecosystem of ponds, lakes, and rivers may result from highly chlorinated flush water arriving over ground or through storm sewers. If the water enters sanitary sewers through sanitary and storm combination systems, the high chlorine might also kill the purifying bacteria in wastewater systems. A dechlorination chemical such as sodium thiosulfate or ascorbic acid should be used prior to or during flushing to inactive the chlorine.

Chlorine Flush

Flush the new main after its 24-hour decontamination period until 1.0 mg/L or less of chlorine is detected. If culture samples are collected before the chlorine is properly flushed from the line, it will interfere with culture results.

First Bacterial Sample Collection

Following a thorough flushing of the chlorinated water, collect one or more 100-mL sample(s) of water for bacterial culture from every 1,200 ft of new main. Also, collect a sample from the end of the line and one sample from each branch. The results of these first samples will help determine if any soil deposited during main installation still remains.

Flush water for a few minutes from copper taps or fire hydrants prior to the collection of the samples. Copper taps are less likely than hydrants to have interfering contaminants. Blow-offs with weep-hole slits cut into them may draw contaminating soil into them. If hydrants are used, open the valve completely to seal off the weep-hole and use the auxiliary valve to control the water flow. Use care to not contaminate the special sterile bacterial sample bottles with soil or with your fingers. Keep samples cool and deliver them to the laboratory.

Main Incubation

The main *must* be allowed to set with the *new* water for at least 24 hours. This important water and pipe contact time must be observed to determine if the pipe wall has been properly decontaminated.

Second Bacterial Sample Collection

After the 24-hour main incubation period, collect a second sample or set of samples from the same sites as the first were collected. These second samples are important in verifying that the wall of the pipe has been properly decontaminated (Figure B-2).

MAIN BREAKS

Broken mains are in jeopardy of soil and sewage contamination that might have infiltrated the surrounding soil. Various disease organisms might accompany the soil and sewage.

To help keep the main clean, efforts should be made to keep soil from entering the pipe. Allow a small amount of water to flow from the break until the pipe has been unearthed. Then valve the water off completely for pipe repairs. This will keep the pipe from drawing large amounts of soil into the distribution system.

Also, the interior of all piping and fittings used in making the repair should be wiped or sprayed with a 1 percent hypochlorite solution.

Thoroughly flush the new pipe and all of the surrounding mains that may have been affected.

Follow up with bacteriological tests in a similar manner as with new main installations.

Bibliography

American Public Health Association (APHA), American Water Works Association (AWWA), and Water Environmental Federation (WEF). 2012. *Standard Methods for the Examination of Water and Wastewater,* 22th ed. Washington, D.C.: APHA.

American Water Works Association (AWWA). *Basic Science Concepts and Applications.* Water Supply Operations series, Part V. Denver, Colo.: AWWA.

_____. Manual M7, *Problem Organisms in Water: Identification and Treatment.* Denver, Colo: AWWA.

_____. Manual M12, *Simplified Procedures for Water Examination.* Denver, Colo.: AWWA.

_____. Manual M48, *Waterborne Pathogens.* Denver, Colo.: AWWA.

_____. Manual M57, *Algae: Source to Treatment.* Denver, Colo.: AWWA.

ANSI/AWWA C651, Disinfecting Water Mains. Denver, Colo.: AWWA.

Battarbee, R.W. 1973. A New Method for the Estimation of Absolute Microfossil Numbers, With Reference Especially to Diatoms, *Limnology and Oceanography,* 18(4):647–653.

BD Difco™ and BD BBL™ Manual, Manual of Microbiological Culture Media, 2nd ed. Sparks, Md.: Difco Laboratories, Div. of Becton Dickinson and Co.

Besner, M.-C., J. Lavoie, C. Morissette, P. Payment, and M. Prevost. 2008. Effect of Water Main Repairs on Water Quality. *Jour. AWWA,* 100(7):95–109.

Brock, T. 1978. Use of Fluorescence Microscopy for Quantifying Phytoplankton, Especially Filamentous Blue-Green Algae. *Limnology and Oceanography,* 23(1):158–160.

Brown, R.A., and D.A. Cornwell. 2007. Using Spore Removal to Monitor Plant Performance for *Cryptosporidium* Removal. *Jour. AWWA,* 99(3):95–109.

Cartier, C., M.C. Besner, B. Barbeau, J. Lavoie, R. Desjardins, and M. Prevost. 2009. Evaluating Aerobic Endospores as Indicators of Intrusion in Distribution Systems. *Jour. AWWA,* 101(7):46–58.

Cornwell, D.A., R.A. Brown, M.J. MacPhee, and R.C. Wichser. 2003. Applying the LT2ESWTR Microbial Toolbox. *Jour. AWWA,* 95(9):76–79.

Cornwell, D.A., M.J. MacPhee, R.A. Brown, and S.H. Via. 2003. Demonstrating *Cryptosporidium* Removal Using Spore Monitoring at Lime-Softening Plants. *Jour. AWWA,* 95(5) 124–133.

Doyle, M., and R.L. Buchanan. 2013. *Food Microbiology Fundamentals and Frontiers,* 4th ed. Washington, D.C.: ASM (American Society for Microbiology) Press.

Dugan, N.R., K.R. Fox, J.H. Owens, and R.J. Miltner. 2001. Controlling *Cryptosporidium* Oocysts Using Conventional Treatment. *Jour. AWWA*, 93(12)64–76.

Hamilton, P.B., M. Proulx, and C. Earle. 2001. Enumerating Phytoplankton with an Upright Compound Microscope Using a Modified Settling Chamber. *Hydrobiologia* 444(1–3):171–175.

Holt, J.G., N.R. Krieg, P.H.A. Sneath, J.T. Staley, and S.T. Williams, eds. 1994. *Bergey's Manual of Determinative Bacteriology*, 9th ed. Philadelphia: Lippincott Williams and Wilkins.

Kerri, K.D. 2008. *Water Treatment Plant Operation*. Vol. 1, 6th ed. Sacramento: California State University School of Engineering.

Lisle, J. 1994. *An Operator's Guide to Bacteriological Testing*. Denver, Colo.: AWWA.

Lund, J.W.G., C. Kipling, and E.D. Le Cren. 1958. The Inverted Microscope Method of Estimating Algal Numbers and the Statistical Basis of Estimations by Counting. *Hydrobiologia*, 11(2):143–170.

Nieminski, E.C., W.D. Bellamy, and L.R. Moss. 2000. Using Surrogates to Improve Plant Performance. *Jour. AWWA*, 92(3):67–78.

Olsen, F.C.W. 1950. Quantitative Estimates of Filamentous Algae. *Transactions of the American Microscopical Society*, 69(3):272–279.

Paxinos, R., and J.G. Mitchell. 2000. A Rapid Utermöhl Method for Estimating Algal Numbers. *Jour. Plankton Res.*, 22(12):2255–2262.

Rice, E.W., K.R. Fox, R.J. Miltner, D.A. Lytle, and C.H. Johnson. 1996. Evaluating Plant Performance With Endospores. *Jour. AWWA*, 88(9):122–130.

Rott, E., N. Salmaso, and E. Hoehn. 2007. Quality Control of Utermöhl-based Phytoplankton Counting and Biovolume Estimates—An easy task or a Gordian Knot? *Hydrobiologia*, 578(1):141–146.

Rott, E. 1981. Some Results From Phytoplankton Counting Intercalibrations. *Schweizerische Zeitschrift für Hydrologie [Aquatic Sciences—Research Across Boundaries]*, 43(1):34–62.

St. Amand, A.L. 2012. 10200 Plankton. In *Standard Methods for the Examination of Water and Wastewater*, 22nd ed., American Public Health Association (APHA), AWWA, and Water Environmental Association. Washington, DC: APHA; 10-2–10-36.

Tille, P. 2013. *Bailey and Scott's Diagnostic Microbiology*, 13th ed. St. Louis, Mo.: Elsevier Mosby.

Versalovic, J. 2011. *Manual of Clinical Microbiology*. 10th ed. Washington, D.C.: ASM Press.

Wehr, J.D., and R.G. Sheath, eds. 2003. *Freshwater Algae of North America: Ecology and Classification*, 2nd ed. San Diego, Calif.: Academic Press, an imprint of Elsevier Science.

ADDITIONAL RESOURCES

Charm Sciences Inc.
36 Franklin Street, Malden, MA 02148-4120
www.charm.com

EMD Chemical Inc.
480 S. Democrat Road, Gibbstown, NJ 08027
www.emdmillipore.com

Fisher Scientific
300 Industry Drive, Pittsburgh, PA 15275
www.fishersci.com

HACH Company
P.O. Box 389, Loveland, CO 80539-0389
www.hach.com

IDEXX Laboratories, Inc.
One IDEXX Drive, Westbrook, ME 04092
www.idexx.com

Krackeler Scientific Inc.
57 Broadway, Albany, NY 12202
www.krackeler.com

Micrology Laboratories
1303 Eisenhower Drive S., Goshen, IN 46526-5360
www.micrologylabs.com

Millipore Corporation
17 Cherry Hill Drive, Danvers, MA 01923
www.millipore.com

Pall Corporation
600 South Wagner Road, Ann Arbor, MI 48103-9019
www.pall.com

Index

NOTE: *f.* indicates figure; *t.* indicates table.

About the Author

DENNIS R. HILL
Microbiologist, Des Moines Water Works

Dennis R. Hill received his microbiology degree from Iowa State University. Prior to his employment at Des Moines Water Works, he worked as a medical center microbiologist for 20 years. In that position, he performed in-depth work relating to several areas of infectious diseases, specializing in anaerobic microbiology, exotic infectious bacteria, mycobacteria, and mycology.

In 1994, he began work as a Microbiologist for Des Moines Water Works. He oversees the microbiology section of the laboratory, has conducted various studies relating to water treatment, and has developed or modified key methods valuable for drinking water laboratories (e.g., aerobic endospore modification and well slide and glycerol phytoplankton methods).

He has written numerous professional papers and given several conference talks on water industry issues. Des Moines Water Works sponsors many tours of its plants, which include his presentation of the microbiology laboratory. His many microbial models and a microscopic pond water show using a large microscope monitor are valued highlights of the tours. He is a member of the American Water Works Association and the American Society for Microbiology.

www.ingramcontent.com/pod-product-compliance
Lightning Source LLC
Chambersburg PA
CBHW042311210326
41598CB00041B/7346